*Sheldon & Company's Text-Books.*

# OLNEY'S HIGHER MATHEMATICS.

***Olney's Introduction to Algebra*** ...............

***Olney's Complete School Algebra*** ..............

***Olney's Key to do. with extra examples*** ...

***Olney's Book of Test Examples in Algebra*** ....

***Olney's University Algebra*** ....................

***Olney's Key to do.*** ....................

***Olney's Elements Geom. & Trigonom.*** (Sch. Ed.)

***Olney's Elements of Geometry.*** Separate........

***Olney's Elements of Trigonometry.*** Separate..

***Olney's Elements of Geometry and Trigonometry.*** (Univ. Ed., with Tables of Logarithms.)......

***Olney's Elements of Geometry and Trigonometry.*** (University Edition, without Tables.).........

***Olney's Tables of Logarithms.*** (Flexible covers.).

***Olney's General Geometry and Calculus*** ......

The universal favor with which these books have been received by educators in all parts of the country, leads the publishers to think that they have supplied a felt want in our educational appliances.

There is one feature which characterizes this series, so unique, and yet so eminently practical, that we feel desirous of calling special attention to it. It is

**The facility with which the books can be used for classes of all grades, and in schools of the widest diversity of purpose.**

Each volume in the series is so constructed that it may be used with equal ease by the youngest and least disciplined who should be pursuing its theme, and by those who in more mature years and with more ample preparation enter upon the study.

*Any of the above sent by mail, post-paid, on receipt of price.*

# TEST EXAMPLES

IN

# ALGEBRA,

ESPECIALLY ADAPTED FOR USE IN CONNECTION WITH
OLNEY'S SCHOOL, OR UNIVERSITY ALGEBRA.

BY
EDWARD OLNEY,
PROFESSOR OF MATHEMATICS, MICHIGAN UNIVERSITY.

NEW YORK:
SHELDON & COMPANY,
No. 8 MURRAY STREET.

# PROF. OLNEY'S MATHEMATICAL WORKS.

---

OLNEY'S COMPLETE SCHOOL ALGEBRA........................

OLNEY'S KEY TO COMPLETE SCHOOL ALGEBRA. ..............

OLNEY'S UNIVERSITY ALGEBRA..............................

OLNEY'S KEY TO UNIVERSITY ALGEBRA.....................

OLNEY'S TEST EXAMPLES IN ALGEBRA......................

OLNEY'S ELEMENTS OF GEOMETRY (Separate)................

OLNEY'S ELEMENTS OF TRIGONOMETRY (Separate)............

OLNEY'S GEOMETRY AND TRIGONOMETRY—One Vol., School Ed

OLNEY'S GEOMETRY AND TRIGONOMETRY—One Vol., Univ. Ed.

OLNEY'S GEOMETRY AND TRIGONOMETRY—One Vol., University Edition, without Tables..............................

OLNEY'S TABLES OF LOGARITHMS (Flexible covers)............

OLNEY'S GENERAL GEOMETRY AND CALCULUS................

---

Entered according to Act of Congress, in the year 1873, by
SHELDON & CO.,
In the Office of the Librarian of Congress, at Washington.

# DUPLICATE EXAMPLES

FOR

## THE CLASS ROOM DRILL.

[NOTE.—The following examples are collected here simply as an aid to the teacher in conducting class room exercises. So far as the principles involved are concerned, they are duplicates of those in the text-book ; but they are so modified in their details, that, if given to the pupils for solution on the blackboard, in class, they will not be going over an exercise which may have been conned by rote. The author recommends that the pupils *study only* the examples in the text-book, but that these here arranged be given for solution in the class, *even to the exclusion* of the others, if thought expedient. They are so numbered by article, page and example, as to render their use practicable with the least possible labor on the part of the teacher.]

### EXAMPLES IN READING LITERAL EXPRESSIONS.

(PAGE 18, ***44.***)

**(1—3.)** $38x^{-5}y^{\frac{3}{5}}$. $c^{\frac{m}{5}-3}$. $13c^{\frac{1}{3}}x^{-4}$. $51m^rn^{\frac{a}{b}}$.

### EXAMPLES IN NOTATION.

(PAGES 24, 25, ***63.***)

**(1.)** Write 3 times the cube root of $x$, added to the square of $a$ minus $b$. $(a-b)^2+3\sqrt[3]{x}$.

(2.) Write the 3rd power of $c$, plus 10 times the product of $a$ cube multiplied by $x$ square, diminished by $b$ times the binomial, the cube root of $x$ minus $y$ square. $c^3 + 10a^3x^2 - b(x^{\frac{1}{3}} - y^2)$.

(3.) Write 6 times $m$ into $n$, minus the fraction $a$ cube, minus $b$ square, divided $m$ plus $n$. $6mn - \frac{a^3 - b^2}{m + n}$.

(4.) Write the fraction, the product of the square root of $x$ plus the cube root of $y$, multiplied by $a$ square minus $b$ cube, divided by the cube root of $m$ plus the 5th root of $n$.

$$\frac{(\sqrt{x} + \sqrt[3]{y})(a^2 - b^3)}{\sqrt[3]{m} + \sqrt[5]{n}}.$$

(5.) Write the fraction, $m$ fifth power increased by 6 times $x$ square $y$ cube, divided by the cube root of the binomial $y$ cube plus $x$ square.

$$\frac{m^5 + 6x^2y^3}{\sqrt[3]{y^3 + x^2}}.$$

(6.) Write the square root of the sum of $a$ square minus $b$ cube equals $x$ minus the square root of $x$ plus $a$.

$$\sqrt{a^2 - b^3} = x - \sqrt{x + a}.$$

(7.) Write one-third of $x$, increased by two, is equal to three $y$ diminished by eleven. $\frac{x}{3} + 2 = 3y - 11$.

(8.) Write $x$ plus the square root of the binomial $a$ square plus $x$ square, equals the fraction, twice $a$ square divided by the square root of the binomial $a$ square minus $x$ square.

$$x + \sqrt{a^2 + x^2} = \frac{2a^2}{\sqrt{a^2 - x^2}}.$$

(9.) The ratio of five $a$ divided by $b$ to $d$ divided by $c$ square, equals the ratio of $x$ square $y$ cube to $y$ square $z$ fourth power.

$$\frac{5a}{b} : \frac{d}{c^2} :: x^2y^3 : y^2z^4.$$

(10.) The cube root of the fraction $a$ plus $b$ divided by $c - d$, is greater than the product of the square of $a - d$ into the third power of $a + b$, less one-fourth of three $x$. $\sqrt[3]{\frac{a + b}{c - d}} > (a - d)^2(a + b)^3 - \frac{3x}{4}$.

## EXERCISES IN READING AND EVALUATING EXPRESSIONS.

(PAGES 25, 26.)

Read the following expressions and find the value of each ; if $a = 6$, $b = 5$, $c = 4$, $d = 1$, and $e = 0$.

(1.) $a^2 \times \overline{a + b} - 2abc = 156.$

(2.) $2a\sqrt{b^2 - ac} + \sqrt{2ac + c^2} = 20.$

(3.) $3a\sqrt{2ac + c^2}$, or $3a(2ac + c^2)^{\frac{1}{2}} = 144.$

(4.) $\sqrt{2a^2 - \sqrt{2ac + c^2}} = 8.$

(5.) $\dfrac{2a + 3c}{6d + 4e} + \dfrac{4bc}{\sqrt{2ac + c^2}} = 14.$

If $a = \frac{1}{2}$, $b = \frac{1}{3}$, $c = \frac{1}{5}$, $d = \frac{2}{7}$, and $x = 2$,

(6.) $(2a + 3b + 5c)\,(8a + 3b - 5c)(2a - 3b + 15c) = 36.$

(7.) $x^3 - \left(\dfrac{1}{a} + \dfrac{1}{b}\right)x^2 + \left(\dfrac{1}{b} - \dfrac{1}{a}\right)x + \dfrac{2}{b^2} = 8.$

(8.) $x^4 - (2a + 3b)x^3 + (3a - 2b)x^2 - cx + bc = 3.$

If $a = b$, and $b = \frac{1}{3}$,

(9.) $\dfrac{5a + b - [3a - (2a - b)]}{a} = 4.$

(10.) $\dfrac{13a + 3b - \{7(a + b) - [3a + 8(a - b)]\}}{2a + 3b} = 1.$

If $b = 3$, $c = 4$, $d = 6$, and $e = 2$,

(11.) $\sqrt{(27b)} - \sqrt[3]{(2c)} + \sqrt{(2e)} = 9.$

(12.) $\sqrt{(3bc)} + \sqrt[3]{(9cd)} - \sqrt[3]{(2e^2)} = 10.$

Let the pupil verify the following expressions, by giving to each letter any value whatever.

(13.) $a(m + n)(m - n) = am^2 - an^2.$

(14.) $\dfrac{x^3 - y^3}{x - y} = x^2 + xy + y^2.$

(**15.**) $(x^4 + x^2 + 1)(x^2 - 1) = x^6 - 1.$

(**16.**) $\frac{a+b}{a-b} - \frac{a-b}{a+b} = \frac{4ab}{a^2 - b^2}.$

---

## ADDITION.

### (Pages 29, 30, *66.*)

(**1.**) Add $7a$, $-3a$, $+8a$, $-a$. *Sum*, $11a$.

(**2.**) Add $10a^2$, $-2a^2$, $-3a^2$, $13a^2$, $-8a^2$. *Sum*, $10a^2$.

(**3.**) Add $mx^{\frac{1}{2}}$, $-2mx^{\frac{1}{2}}$, $-3mx^{\frac{1}{2}}$, $7mx^{\frac{1}{2}}$, $-4mx^{\frac{1}{2}}$. *Sum*, $-mx^{\frac{1}{2}}$.

(**4.**) Add $6cy$, $-2cy$, $4cy$, $-cy$, $8cy$, $-10cy$. *Sum*, $5cy$.

(**5.**) Add $2nx^2$, $-6nx^2$, $-10nx^2$, $7nx^2$, $7nx^2$. *Sum*, 0.

(**6.**) Add $3cx^3$, $-8cx^3$, $2cx^3$, $-12cx^3$, $-cx^3$. *Sum*, $-16cx^3$.

(**7.**) Add $3y^{\frac{1}{5}}$, $2y^{\frac{1}{5}}$, $-5y^{\frac{1}{5}}$, $-11y^{\frac{1}{5}}$, $8y^{\frac{1}{5}}$. *Sum*, $-3y^{\frac{1}{5}}$.

(**8.**) Add $-6x^3$, $-2x^3$, $-5x^3$, $-4x^3$. *Sum*, $-17x^3$.

(**9.**) Add $-3n\sqrt[3]{y}$, $2n\sqrt[3]{y}$, $-5n\sqrt[3]{y}$, $12n\sqrt[3]{y}$. *Sum*, $6n\sqrt[3]{y}$.

(**10.**) Add $5m\sqrt{n}$, $-2mn^{\frac{1}{2}}$, $-3m\sqrt{n}$, $4mn^{\frac{1}{2}}$. *Sum*, $4mn^{\frac{1}{2}}$.

(**11.**) Add $6m^{\frac{5}{3}}x^{\frac{1}{4}}$, $2m^{\frac{5}{3}}\sqrt[4]{x}$, $-2m^{\frac{5}{3}}\sqrt[4]{x}$, $m^{\frac{5}{3}}x^{\frac{1}{4}}$. *Sum*, $7m^{\frac{5}{3}}x^{\frac{1}{4}}$.

(**12.**) Add $\frac{2}{3}xy$, $-\frac{1}{3}xy$, $\frac{2}{3}xy$. *Sum*, $xy$.

(**13.**) Add $15x^2$, $-10x^2$, $x^2$, $-18x^2$. *Sum*, $-12x^2$.

(**14.**) Add $2a^2$, $-1\frac{1}{4}a^2$, $-\frac{1}{3}a^2$, $\frac{1}{6}a^2$. *Sum*, $\frac{7}{12}a^2$.

(**15.**) Add $2\sqrt[3]{m}$, $-3\sqrt[3]{m}$, $7\sqrt[3]{m}$, $4\sqrt[3]{m}$. *Sum*, $10\sqrt[3]{m}$.

(**16.**) Add $3y$, $-\frac{3}{5}y$, $-4y$, $-\frac{1}{3}y$, $-\frac{1}{6}y$. *Sum*, $-2\frac{1}{10}y$.

### (Page 30, *67.*)

(**17.**) A boy drawing his sleigh up a high and slippery hill, ascends 10 rods, slips back 3, ascends 7 rods, slips back 2, ascends 8 rods, slips back 3. How far does he ascend; or, what is the *sum* of his movements? *Ans.*, 17 rods.

(**18.**) A certain quality of sugar fluctuates in price as follows during a year: the price rises 4 cents, then 2 cents, then falls 1 cent, then rises 3 cents, then falls 5 cents. Was it, on the whole, higher or lower at the end of the year than at the beginning; *i. e.*, what was the *sum* of the fluctuations? *Ans.*, It rises 3 cents. $(+3.)$

(**19.**) During a series of years a man accumulates in money and property, \$11$m$, \$5$m$, and \$6$m$; then loses \$3$m$, \$12$m$, and \$$m$; then accumulates \$10$m$; and then loses \$16$m$. What is the *sum* of his efforts? *Ans.*, 0. He neither gains nor loses.

## (Page 31, 70.)

(**1.**) Add $3xy$, $-2bc$, $2x$, $-m$. *Sum*, $3xy - 2bc + 2x - m$.

(**2.**) Add $4cd$, $2mn$, $-5y^{\frac{1}{2}}$, $z^2$, $-1$.
*Sum*, $4cd + 2mn - 5y^{\frac{1}{2}} + z^2 - 1$.

(**3.**) Add $-3\sqrt{x}$, $2\sqrt{y}$, $-5x$, $3y$.
*Sum*, $2\sqrt{y} - 3\sqrt{x} - 5x + 3y$.

(**4.**) Add $2x^{-1}m$, $-12y^{-2}z$, $-16y^{\frac{3}{4}}$, and find the value, if $x=2$, $m=\frac{1}{2}$, $y=1$, and $z=0$. *Sum*, $-$

## (Page 33, 72.)

(**1.**) Add $12ax - 2cy + z$, $10ax + 4cy - 5z + 1$, $-ax - 3cy - m + 6$, $-8ax - 5z + 3m$. *Sum*, $13ax - cy - 9z + 2m + 7$.

(**2.**) Add $2a - 3x^2$, $-7a + 5x^2$, $-3a + x^2$, $a - 3x^2$. *Sum*, $-7a$.

(**3.**) Add $2xy - 2x^2$, $3x^2 + xy$, $x^2 + xy$, $4x^2 - 3xy$. *Sum*, $6x^2 + xy$.

(**4.**) Add $8a^2x^2 - 3ax$, $7ax - 5xy$, $9xy - 5ax$, $2a^2x^2 + xy$.
*Sum*, $10a^2x^2 + 5xy - ax$.

(**5.**) Add $10b^2 - 3a^2x - b^2 + 2a^2x^2$, $50 + 2a^2x$, $a^2x^2 + 120$.
*Sum*, $9b^2 + 3a^2x^2 - a^2x + 170$.

(**6.**) Add $5x - 3a + b + 7$ and $-4a - 3x + 2b - 9$.
*Sum*, $2x - 7a + 3b - 2$.

(**7.**) Add $2a + 3b - 4c - 9$ and $5a - 3b + 2c - 10$.
*Sum*, $7a - 2c - 19$.

(**8.**) Add $3a + 2b - 5$, $a + 5b - c$, and $6a - 2c + 3$.
*Sum*, $10a + 7b - 3c - 2$.

(**9.**) Add $3a^2 + 4bc - e^2 + 10$, $-5a^2 + 6bc + 2e^2 - 15$, and $-4a^2 - 9bc - 10e^2 + 21$. *Sum*, $bc - 6a^2 - 9e^2 + 16$.

**(10.)** Add $6xy - 12x^2$, $-4x^2 + 3xy$, $4x^2 - 2xy$, and $-3xy + 4x^2$.
*Sum*, $4xy - 8x^2$.

**(11.)** Add $4ax - 130 + 3x^{\frac{1}{2}}$, $5x^2 + 3ax + 9x^2$, $7xy - 4x^{\frac{1}{2}} + 90$, and $\sqrt{x} + 40 - 6x^2$.
*Sum*, $7ax + 8x^2 + 7xy$.

**(12.)** Add $2a^2 - 3ab + 2b^3 - 3a^2$, $3b^3 - 2a^2 + a^3 - 5c^3$, $4c^3 - 2b^3 + 5ab + 100$, and $20ab + 16a^2 - bc - 20$.
*Sum*, $13a^2 + 2ab + 3b^3 + a^3 - c^3 - 20 - bc$.

**(13.)** Add $2x + 3a$, $4x + a$, $5x + 8a$, $7x + 2a$, and $x + a$.
*Sum*, $19x + 15a$.

**(14.)** Add $7x^2 - 5bc$, $3x^2 - bc$, $x^2 - 4bc$, $5x^2 - bc$, and $4x^2 - 4bc$.
*Sum*, $20x^2 - 15bc$.

**(15.)** Add $4ax + 3by$, $5ax + 8by$, $8ax + 6by$, and $20ax + by$.
*Sum*, $37ax + 18by$.

**(16.)** Add $10cz - 2ax^2$, $15cz - 3ax^2$, $24cz - 9ax^2$, and $3cz - 8ax^2$.
*Sum*, $52cz - 22ax^2$.

**(17.)** Find the sum of $3x^2y^2 - 10y^4$, $-x^2y^2 + 5y^4$, $8x^2y^2 - 6y^4$, and $4x^2y^2 + 2y^4$.
*Sum*, $14x^2y^2 - 9y^4$.

**(18.)** Add $a + b + c + d$, $a + b + c - d$, $a + b - c + d$, $a - b + c + d$, and $-a + b + c + d$.
*Sum*, $3a + 3b + 3c + 3d$.

**(19.)** Required the sum of $10a^2b - 12a^3bc - 15b^2c^4 + 10$, $-4a^2b + 8a^3bc - 10b^2c^4 - 4$, $-3a^2b - 3a^3bc + 20b^2c^4 - 3$, and $2a^2b + 12a^3bc + 5b^2c^4 + 2$.
*Sum*, $5a^2b + 5a^3bc + 5$.

**(20.)** Add $a^2 + b^2 + c^2 + d^2$, $ab - 2a^2 + ac - 2c^2 + ad - 2d^2$, $a^3 - 3ab + b^3 - 3ac + c^3 - 3ad$, and $2ab - a + 2ac - b + 2ad - c$.
*Sum*, $a^3 + b^3 + c^3 + b^2 - a^2 - c^2 - d^2 - a - b - c$.

**(21.)** Add $a^m - b^n + 3x^p$, $2a^m - 3b^n - x^p$, and $a^m + 4b^n - x^q$.
*Sum*, $4a^m + 2x^p - x^q$.

**(22.)** Add $30 - 13x^{\frac{1}{2}} - 3xy$, $23 - 10x^{\frac{1}{2}} - 4xy$, $-14x^{\frac{1}{2}} - 7xy + 14$, $-5xy + 10 - 16^{\frac{1}{2}}$, and $1 - 2x^{\frac{1}{2}} - xy$.
*Sum*, $78 - 55x^{\frac{1}{2}} - 20xy$.

**(23.)** Add $3x^3 + 4x^2 - x$, $2x^3 + x^2 - 3x$, $7x^3 + 2x^2 - 2x$, and $4x^3 + 2x^2 - 3x$.
*Sum*, $16x^3 + 9x^2 - 9x$.

**(24.)** Add $7a^3 - 3a^2b + 2ab^2 - 3b^3$, $ab^2 - a^2b - b^3 + 4a^3$, $-5b^3 + 5ab^2 - 4a^2b + 6a^3$, and $-a^2b + 4ab^2 - 4b^3 + a^3$.
*Sum*, $18a^3 - 9a^2b + 12ab^2 - 13b^3$.

(**25.**) Add $2x^2y - x + 2$, $x^2y - 4x + 3$, $4x^2y - 3x + 1$, and $5x^2y - 7x + 7$. *Sum*, $12x^2y - 15x + 13$.

(**26.**) Add $2x^{\frac{1}{2}} - 10y$, $3\sqrt{xy} + 10x$, $2x^2y + 25y$, $12xy - \sqrt{xy}$, $-8y + 17\sqrt{xy}$. *Sum*, $2x^2y + 12xy + 10x + 2x^{\frac{1}{2}} + 19\sqrt{xy} + 7y$.

(**27.**) Add $2bx - 12$, $3x^2 - 2bx$, $5x^2 - 3\sqrt{x}$, $3\sqrt{x} + 12$, $x^2 + 3$.
*Sum*, $9x^2 + 3$.

(**28.**) Add $10b^2 - 3bx^2$, $2b^2x - b^2$, $10 - 2bx^2$, $b^2x - 20$, $3bx^2 + b^2$.
*Sum*, $10b^2 + 3b^2x - 2bx^2 - 10$.

(**29.**) Add $2a^2 - 3ax^{\frac{1}{2}} + x^2$, $2ax^{\frac{1}{2}} - 13xy + 8$, $10a^2 - xy - 4$.
*Sum*, $12a^2 - ax^{\frac{1}{2}} + x^2 - 14xy + 4$.

(**30.**) Add $x^5 + 5x^4y + 10x^3y^2 + 10x^2y^3 + 5xy^4 + y^5$, and $y^5 - 5y^4x + 10y^3x^2 - 10y^2x^3 + 5yx^4 - x^5$. *Sum*, $2y^5 + 20y^3x^2 + 10yx^4$.

(**31.**) Add $ax + 2by + cz$, $\sqrt{x} + \sqrt{y} + \sqrt{z}$, $3y^{\frac{1}{2}} - 2x^{\frac{1}{2}} + 3z^{\frac{1}{2}}$, $4cz - 3ax - 2by$, $2ax - 4\sqrt{y} - 2z^{\frac{1}{2}}$. *Sum*, $5cz - \sqrt{x} + 2\sqrt{z}$.

(**32.**) Add $\sqrt{a} - 3xy - \sqrt[3]{m} - n + 6xy + 5\sqrt{a} + 7n - 7xy + 9\sqrt{a} - 7\sqrt[3]{m} + 16n - 5\sqrt{a}$. *Sum*, $10\sqrt{a} - 4xy - 8\sqrt[3]{m} + 22n$.

## (PAGE 37, *73.*)

(**1.**) Add $cy^{\frac{1}{2}}$, $-ay^{\frac{1}{2}}$, $-by^{\frac{1}{2}}$, and $3my^{\frac{1}{2}}$, with respect to $y^{\frac{1}{2}}$.
*Sum*, $(c - a - b + 3m)y^{\frac{1}{2}}$.

(**2.**) Add $2axy$, $-3bxy$, $4cxy$, and $-dxy$, with respect to $xy$.
*Sum*, $(2a - 3b + 4c - d)xy$.

(**3.**) Add $2my - 2cx + 1$, $3dy + 4cx + 5$, $12my - 3nx - 4$, and $2dy - nx + 3$ with respect to $x$ and $y$.
*Sum*, $(14m + d)y + (2c - 4n)x + 5$.

(**4.**) Add $x^3 + ax^2 + bx + 2$, and $x^3 + cx^2 + dx - 1$, with respect to $x$. *Sum*, $2x^3 + (a + c)x^2 + (b + d)x + 1$.

(**5.**) Add $ax^4 - bx^3 + cx^2$, $bcx^2 - acx^3 - c^2x$, and $ax^2 + c - bx$, with respect to $x$. *Sum*, $ax^4 - (b + ac)x^3 + (c + bc + a)x^2 - (c^2 + b)x + c$.

(**6.**) Add $(2m - 3n + 1)y^{\frac{1}{3}}$, $(3m + 4n - 6)y^{\frac{1}{3}}$, and $(2 - m)y^{\frac{1}{3}}$.
*Sum*, $(4m + n - 3)\sqrt[3]{y}$.

(7.) Add $2a\sqrt{x} + 3by^{-\frac{1}{2}} - ax^{\frac{1}{2}} + 2my^{-\frac{1}{2}} - 3a\sqrt{x} + \frac{2b}{\sqrt{y}} - ax^{\frac{1}{2}}$

*Sum*, $(5b + 2m)y^{-\frac{1}{2}} - 3a\sqrt{x}$, or $\frac{5b + 2m}{\sqrt{y}} - 3a\sqrt{x}$.

(8.) Add $(a + b)x + (c - d)y - x\sqrt{2}$, $(a - b)x + (3c + 2d)y + 5x\sqrt{2}$, $2bx + 3dy - 2x\sqrt{2}$, and $-3bx - dy - 4x\sqrt{2}$.

*Sum*, $(2a - b - 2\sqrt{2})x + (4c + 3d)y$.

## (Page 38, 74.)

(1.) Add $3(x^2 - y^2)$, $8(x^2 + y^2)$, and $-5(x^2 - y^2)$.

*Sum*, $6(x^2 - y^2)$.

(2.) Add $17a(x + 3ay) + 12a^3b^4c^2$, $8 - 18ay - 8a^3b^4c^2$, $-7a(x + 3ay) - 4 + 17ay$. *Sum*, $10a(x + 3ay) + 4a^3b^4c^2 - ay + 4$.

(3.) Add $3(x + y)^2 - 4(a - b)^3$, $(x + y)^2 - (a - b)^3$, $-7(a - b)^3 + 5(x + y)^2$, and $2(x + y)^2 - (a - b)^3$. *Sum*, $11(x + y)^2 - 13(a - b)^3$.

(4.) Add $8ax + 2(x + a) + 3b$, $9ax + 6(x + a) - 9b$, and $11x + 6b - 7ax - 8(x + a)$. *Sum*, $10ax + 11x$.

(5.) Add $(a + b)\sqrt{x}$ and $(c + 2a - b)\sqrt{x}$. *Sum*, $(c + 3a)\sqrt{x}$.

(6.) Add $28a^3(x + 5y) + 21$, $18a - 13a^3(x + 5y)$, $-15a^3(x + 5y) - 8$.

*Sum*, $18a + 13$.

(7.) Add $\frac{5a}{b} - \frac{3c^2}{a} + \frac{7\sqrt{bc}}{x} - 9\left(\frac{ab + x}{d}\right)$ and $\frac{8a}{b} + \frac{7c^2}{a} - 12\frac{\sqrt{bc}}{x} + 6\left(\frac{ab + x}{d}\right)$. *Sum*, $\frac{13a}{b} + \frac{4c^2}{a} - 5\frac{\sqrt{bc}}{x} - 3\left(\frac{ab + x}{d}\right)$.

(8.) Add $\frac{4a}{y} - \frac{3m^3}{c} + \frac{6\sqrt[3]{p}}{z} - \frac{3(q + r)}{s}$, and $\frac{9a}{y} + \frac{8m^3}{c} - \frac{12\sqrt[3]{p}}{z} + \frac{4(q + r)}{s}$. *Sum*, $\frac{13a}{y} + \frac{5m^3}{c} - \frac{6\sqrt[3]{p}}{z} + \frac{q + r}{s}$.

---

# SUBTRACTION.

## (Page 42, 77.)

(1.) From $24ab + 7cd$ subtract $18ab + 7cd$. *Rem.*, $6ab$.

(2.) Subtract $7a - 5b + 3ax$ from $12a + 10b + 13ax - 3ab$.

*Rem.*, $5a + 15b + 10ax - 3ab$.

**(3.)** From $3ab - 7ax + 7ab + 3ax$ take $4ab - 3ax - 4xy$.
*Rem.*, $6ab - ax + 4xy$.

**(4.)** From $a - b$, subtract $a + b$. *Rem.*, $-2b$.

**(5.)** From $7xy - 5y + 3x$, subtract $3xy + 3y + 3x$.
*Rem.*, $4xy - 8y$.

**(6.)** What is the difference between $7ax^2 + 5xy - 12ay + 5bc$, and $4ax^2 + 5xy - 8ay - 4cd$. *Ans.*, $3ax^2 - 4ay + 5bc + 4cd$.

**(7.)** From $8x^2 - 3ax + 5$, take $5x^2 + 2ax + 5$.
*Rem.*, $3x^2 - 5ax$.

**(8.)** From $a + b + c$, take $-a - b - c$. *Rem.*, $2a + 2b + 2c$.

**(9.)** From $17cx + 12px - 7ny^2$ take $14cx - 7px + 3ny^2 - b$.
*Rem.*, $3cx + 19px - 10ny^2 + b$.

**(10.)** From $13xy - 14xy^{\frac{1}{2}} + 17ay$ take $4xy^{\frac{1}{2}} + 7xy - 10ay$.
*Rem.*, $6xy - 18xy^{\frac{1}{2}} + 27ay$.

**(11.)** From $9a^2x^2 - 16 + 10ay^5$ take $4a^2x^2 - 8 + 7ay^5 + y$.
*Rem.*, $5a^2x^2 - 8 + 3ay^5 - y$.

**(12.)** From $4a^m + 2x^p - x^q$ take $a^m - b^n + 3x^p$ and $2a^m - 3b^n - x^p$.
*Rem.*, $a^m + 4b^n - x^q$.

**(13.)** From $\sqrt{x^2 + y^2} + 4(x + y) - 3\sqrt{a + x}$ take $3(x + y) - 2(x^2 + y^2)^{\frac{1}{2}} + 3(a + x)^{\frac{1}{2}}$. *Rem.*, $3\sqrt{x^2+y^2} + (x + y) - 6(a+x)^{\frac{1}{2}}$.

**(14.)** From $4y^2 + 4xy + x^2 - 2a(x + y) + 6\sqrt{a^2 - x^2} - 8\sqrt[3]{b^2 - y^2}$ take $4x^2 + 4xy + y^2 - 4a(x + y) - 10\sqrt[3]{b^2 - y^2} + 4\sqrt{a^2 - x^2}$.
*Rem.*, $3(y^2 - x^2) + 2a(x + y) + 2\sqrt{a^2 - x^2} + 2\sqrt[3]{b^2 - y^2}$.

**(15.)** From $(a + m - n)\sqrt{x}$ take $(a + m + n)\sqrt{x} - 70$.
*Rem.*, $70 - 2n\sqrt{x}$.

**(16.)** From $(x - y)\sqrt{a - b} + a^2xy$ take $(y - x)\sqrt{a - b} + b^2xy$.
*Rem.*, $2(x - y)\sqrt{a - b} + (a^2 - b^2)xy$.

**(17.)** From the sum of $3x^3 - 4ax + 3y^2$, $4y^2 + 5ax - x^3$, $y^2 - ax + 5x^3$, and $3ax - 2x^2 - y^2$, take the sum of $5y^2 - x^2 + x^3$, $ax - x^3 + 4x^2$, $3x^3 - ax - 3y^2$, and $7y^2 - ax + 7$.
*Result*, $4x^3 + 4ax - 2y^2 - 5x^2 - 7$

(**18.**) From the sum of $x^2y^2 - x^2y - 3xy^2$, $9xy^2 - 15 - 3x^2y^2$, and $70 + 2x^2y^2 - 3x^2y$, subtract the sum of $5x^2y^2 - 20 + xy^2$, $3x^2y - x^2y^2 + ax$, and $3xy^2 - 4x^2y^2 - 9 + a^2x^2$.

*Result*, $2xy^2 - 7x^2y - ax - a^2x^2 + 84$.

(**19.**) From $a^3x^2y^2 - m^2x^3 + 3cx - 4x^2 - 9$ take $a^2x^2y^2 - n^2x^3 + c^2x + bx^2 + 3$.

*Result*, $(a^3 - a^2)x^2y^2 - (m^2 - n^2)x^3 + (3c - c^2)x - (4 + b)$.

(**20.**) From the sum of $6x^2y - 11ax^3$ and $8x^2y + 3ax^3$, take $4x^2y - 4ax^3 + a$. *Result*, $10x^2y - 4ax^3 - a$.

(**21.**) From the sum of $15a^2b + 8cdx - 3$ and $24 - 8a^2b + 2cdx$, take the sum of $12a^2b - 3cdx - 8$ and $16 + cdx - 4a^2b$.

*Result*, $12cdx + 13 - a^2b$.

(**22.**) From the difference between $8ab - 12cy$ and $-3ab + 4cy$, take the sum of $5ab - 7cy$ and $ab + cy$. *Result*, $5ab - 10cy$.

(**23.**) From $-17x^3 + 9ax^2 - 7a^2x + 15a^3$ take $-19x^3 + 9ax^2 - 9a^2x + 17a^3$. *Rem.*, $2x^3 + 2a^2x - 2a^3$.

(**24.**) From $x^3 + 3x^2 + 3x + 1$ take $x^3 - 3x^2 + 3x - 1$.

*Rem.*, $6x^2 + 2$.

(**25.**) From $9a^mx^2 - 13 + 20ab^3x - 4b^mcx^2$ take $3b^mcx^2 + 9a^mx^2 - 6 + 3ab^3x$. *Rem.*, $17ab^3x - 7b^mcx^2 - 7$.

(**26.**) From $a - x - (x - 2a) + 2a - x$ take $a - 2x - (2a - x) + (x - 2a)$. *Rem.*, $8a - 3x$.

(**27.**) Take $3bxy - (5 + c)z^2 - (3a - 2b)(x + y)^{\frac{1}{2}}$ from $(4a - 5b)\sqrt{x + y} + (a - b)xy - cz^2$.

*Rem.*, $(7a - 7b)\sqrt{x + y} + (a - 4b)xy + 5z^2$.

## (Page 46, 78.)

(**28.**) Show that $3ax - 5by^{\frac{1}{2}} + m - (2ax - 4by^{\frac{1}{2}} + m) = ax - by^{\frac{1}{2}}$.

(**29.**) Show that $3x - 2y - 5m - (-2x + 4y - 6m) = 5x - 6y + m$.

(**30.**) Remove the parenthesis from $3cd - (2cd - 4ax) + ax + (-5ax - cd)$. *Result*, 0.

## (Page 47, 79.)

(**31.**) Introduce the 2nd, 3rd, and 4th terms of the following expression within a parenthesis: $5mx - 2cd - 4y + 1 - 3nz$.

*Result*, $5mx - (2cd + 4y - 1) - 3nz$.

**(32.)** Include in brackets the 2nd and 3rd, and the 5th and 6th terms of the following polynomial: $2c^2 - 2a - 1 + 5m + xy - 4n$

*Result*, $2c^2 - (2a + 1) + 5m + (xy - 4n)$.

**(33.)** Show that $5c - 2xy - 4b + 5 = 5c - 2xy - (4b - 5) = 5c - (2xy + 4b - 5)$.

**(34.)** Show that $a^2b + xy - 7am - mx - 6 + 13x^2 = a^2b + xy - 7am - (mx + 6 - 13x^2) = a^2b + xy - (7am + mx + 6 - 13x^2) = a^2b - (- xy + 7am + mx + 6 - 13x^2) = a^2b + xy - 7am - mx - (6 - 13x^2)$.

**(35.)** $(1 - 2x + 3x^2) + (3 + 2x - x^2) = 4 + 2x^2$.

**(36.)** $5a - 4b + 3c + (- 3a + 2b - c) = 2a - 2b + 2c$.

**(37.)** $(a - b - c) + (b + c - d) + (d - e + f) + (e - f - g) = a - g$.

**(38.)** $3(x^2 + y^2) - \{(x^2 + 2xy + y^2) - (2xy - x^2 - y^2)\} = x^2 + y^2$.

**(39.)** $\frac{1}{2}(2x - 4y) - \frac{1}{3}(16x - 15y) = 3y - 4\frac{1}{3}x$.

**(40.)** $\frac{1}{2}(a + b + c) - \frac{1}{2}(a - b - c) = b + c$.

**(41.)** $\frac{1}{3}(9x + 6y) - \frac{1}{5}(10x - 20y) + x - y = 2x + 5y$.

**(42.)** $2ax - 3by - (\frac{1}{2}ax + \frac{1}{3}by) - \frac{1}{2}(- ax - by) + (- 2ax + \frac{1}{3}by) - 2\frac{1}{2}by$.

(PAGE 48, ***80.***)

**(43.)** $a - (x - a) - \{x - (a - x)\} = 3a - 3x$.

**(44.)** $1 - \{1 - [1 - (1 - x)]\} = x$.

**(45.)** $a - (b - c) - (a - c) + c - (a - b) = 3c - a$.

**(46.)** $a - \{a + b - [a + b + c - (a + b + c + d)]\} = - b - d$.

**(47.)** From $my - cz + 2x$ take $ny + z - bx$.

*Rem.*, $(m - n)y - (c + 1)z + (2 + b)x$.

**(48.)** From $ax^2 - bxy + cy^2$ take $a_1x^2 - b_1xy + c_1y^2$.

*Rem.*, $(a - a_1)x^2 + (b_1 - b)xy + (c - c_1)y^2$.

**(49.)** From $(2a - 5b)\sqrt{x + y} + (a - b)xy - cz^2$ take $3bxy - (5 + c)z^2 - (3a - b)(x + y)^{\frac{1}{2}}$.

*Rem.*, $(5a - 6b)\sqrt{x + y} + (a - 4b)xy + 5z^2$.

**(50.)** From $2x - y + (y - 2x) - (x - 2y)$ take $y - 2x - (2y - x) + (x + 2y)$. *Rem.*, $y - x$.

(**51.**) From $(a+b)(x+y)-(c-d)(x-y)+h^2$ take $(a-b)(x+y)+(c+d)(x-y)+k^2$. *Rem.*, $2b(x+y)-2c(x-y)+h^2-k^2$.

---

## MULTIPLICATION.

### (Page 54, *90.*)

Prove the following :

(**1.**) $16^{\frac{3}{2}} \times 16^{\frac{3}{4}} = 16^{\frac{18}{8}} = 16^{\frac{9}{4}}$.

(**2.**) $64^{-\frac{1}{2}} + 64^{-\frac{1}{3}} = 64^{-\frac{5}{6}}$.

(**3.**) $36^{-\frac{1}{2}} \times 36^{\frac{1}{2}} = \frac{1}{\sqrt{36}} \times \sqrt{36} = \frac{1}{6} \times 6 = 1$.

For further examples see page 65 in Division, in the text-book. Assign the quotient and divisor.

### (Page 54, *91.*)

(**1.**) $2a^2xy \times 5axy^3 \times 6a^3x^2 = 60a^6x^4y^4$.

(**2.**) $10a^2m \times 5am \times 3a^3m^2y = 150a^6m^4y$.

(**3.**) $3mn \times 2m^2y \times -4m \times -5y^3 \times -my = -120m^5ny^5$.

(**4.**) $2a^nx^2y \times -3a^mxy^3 \times -5ax^ry = 30a^{m+n+1}x^{3+r}y^5$.

(**5.**) $-5x^my^n \times -2a^2x^ny^m \times -3axy = -30a^3x^{m+n+1}y^{m+n+1}$.

For further examples see page 68 of text-book. Assign quotient and divisor.

(**6.**) $x^{\frac{1}{3}} \times x^{-\frac{1}{3}} = x^0$. Also $x^{\frac{1}{3}} \times x^{-\frac{1}{3}} = \frac{x^{\frac{1}{3}}}{x^{\frac{1}{3}}} = 1$. $\therefore$ $x^0 = 1$.

(**7.**) $\sqrt{a} \times \sqrt[3]{a} = a^{\frac{1}{2}} \times a^{\frac{1}{3}} = a^{\frac{5}{6}} = \sqrt[6]{a^5}$.

(**8.**) $-3\sqrt{a^3} \times 2\sqrt[3]{a^2} = -3a^{\frac{3}{2}} \times -2a^{\frac{2}{3}} = 6a^{\frac{13}{6}} = 6\sqrt[6]{a^{13}}$.

(**9.**) $2\sqrt[5]{x^2} \times 3\sqrt[3]{x^5} = 6x^{\frac{31}{15}} = 6\sqrt[15]{a^{31}}$.

(**10.**) $a^{-m}b^{-n} \times a^mb^n = 1$.

### (Page 56, *92.*)

See pages 69, 70–75 of text-book, assigning as directed above.

### (PAGE 59, *94.*)

(**1.**) $(3x^3 + 2y)^2 = 9x^6 + 12x^3y + 4y^2.$

(**2.**) $(m^{\frac{2}{3}} + n^{\frac{1}{2}})^2 = m^{\frac{4}{3}} + 2m^{\frac{2}{3}}n^{\frac{1}{2}} + n.$

(**3.**) $(a^{-2} + 3x^2)^2 = a^{-4} + 6a^{-2}x^2 + 9x^4.$

(**4.**) $(x^3 + x)^2 = x^6 + 2x^4 + x^2.$

(**5.**) $(\frac{1}{3}a^{-\frac{1}{2}}b^{-\frac{1}{3}} + \frac{1}{2}a^{\frac{1}{2}}b^{\frac{1}{3}})^2 = \frac{1}{9}a^{-1}b^{-\frac{2}{3}} + \frac{1}{3} + \frac{1}{4}ab^{\frac{2}{3}}.$

### (PAGE 60, *95.*)

(**1.**) $(3x - 2y)^2 = 9x^2 - 12xy + 4y^2.$

(**2.**) $(2m^{-\frac{1}{2}} - 3n^{\frac{1}{3}})^2 = 4m^{-1} - 12m^{-\frac{1}{2}}n^{\frac{1}{3}} + 9n^{\frac{2}{3}}.$

(**3.**) $(3x^{\frac{m}{n}} - 5y^{-2})^2 = 9x^{\frac{2m}{n}} - 30x^{\frac{m}{n}}y^{-2} + 25y^{-4}.$

(**4.**) $(a^{-r} - b^{-s})^2 = a^{-2r} - 2a^{-r}b^{-s} + b^{-2s}.$

(**5.**) $(4x^{-\frac{1}{n}} - 3x^{-\frac{1}{m}})^2 = 16x^{-\frac{2}{n}} - 24x^{-\frac{m+n}{mn}} + 9x^{-\frac{2}{m}}.$

### (PAGE 61, *96.*)

(**1.**) $(6a + b) \times (6a - b) = 36a^2 - b^2.$

(**2.**) $(2y - 3x) \times (2y + 3x) = 4y^2 - 9x^2.$

(**3.**) $(\frac{1}{2}x - \frac{1}{3}y^2) \times (\frac{1}{2}x + \frac{1}{3}y^2) = \frac{1}{4}x^2 - \frac{1}{9}y^4.$

(**4.**) $(3a^2 - 2b^3) \times (3a^2 + 2b^3) = 9a^4 - 4b^6.$

(**5.**) $(3 - x^n) \times (3 + x^n) = 9 - x^{2n}.$

(**6.**) $(a^{\frac{1}{2}} - b^{\frac{1}{2}}) \times (a^{\frac{1}{2}} + b^{\frac{1}{2}}) = a - b.$

(**7.**) $(x^{\frac{2}{3}} - y^{\frac{2}{3}}) \times (x^{\frac{2}{3}} + y^{\frac{2}{3}}) = x^{\frac{4}{3}} - y^{\frac{4}{3}}.$

(**8.**) $(2a^{\frac{1}{2}}b^{\frac{1}{3}} - 3a^{\frac{1}{3}}b^{\frac{1}{2}}) \times (2a^{\frac{1}{2}}b^{\frac{1}{3}} + 3a^{\frac{1}{3}}b^{\frac{1}{2}}) = 4ab^{\frac{2}{3}} - 9a^{\frac{2}{3}}b.$

## DIVISION.

### (PAGE 65, *107.*)

See examples in text-book, pages 55 and 56, from *Ex.* 6 to 20 in Multiplication. Assign the product and either factor.

### (PAGE 67, *111.*)

(**1.**) Divide $36x^2y^2$ by $9xy$. *Quot.*, $4xy$.

(**2.**) Divide $30a^2by^2$ by $-6aby$. *Quot.*, $-5ay$.

(**3.**) Divide $-42c^3x^3y$ by $7c^2x^2$. *Quot.*, $-6cxy$.

(**4.**) Divide $-4ax^2y^3$ by $-axy^2$. *Quot.*, $+4xy$.

(**5.**) Divide $16a^5b^3cx$ by $-4a^3bdy$. *Quot.*, $-\frac{4a^2b^2cx}{dy}$.

(**6.**) Divide $-18a^3b^2c^2$ by $12a^5b^3x$. *Quot.*, $-\frac{3c^2}{2a^2bx}$.

(**7.**) Divide $17xyzw^2$ by $xzyw$. *Quot.*, $17w$.

(**8.**) Divide $-12a^3b^3c^3$ by $-6abc$. *Quot.*, $2a^2b^2c^2$.

(**9.**) Divide $6a^{\frac{4}{5}}$ by $3a^{\frac{2}{3}}$. *Quot.*, $2a^{\frac{2}{15}}$.

(**10.**) Divide $32a^{\frac{1}{2}}b^{\frac{1}{3}}$ by $4a^{\frac{1}{4}}b^{\frac{1}{6}}$. *Quot.*, $8a^{\frac{1}{4}}b^{\frac{1}{6}}$.

(**11.**) Divide $144ab^2$ by $24a^{\frac{1}{2}}b^{\frac{1}{4}}$. *Quot.*, $6a^{\frac{1}{2}}b^{\frac{7}{4}}$.

(**12.**) Divide $mx^{\frac{5}{7}}$ by $nx^{\frac{2}{3}}$. *Quot.*, $\frac{m}{n}x^{\frac{1}{21}}$.

(**13.**) Divide $3x^m$ by $2x^n$. *Quot.*, $\frac{3}{2}x^{m-n}$.

(**14.**) $15a^{-2}b^3 \div 3a^2b^{-3} = 5a^{-4}b^6$.

(**15.**) $20a^{\frac{1}{2}}b^{\frac{1}{3}}x \div 10ab^{\frac{1}{6}}x^{\frac{1}{2}} = 2a^{-\frac{1}{2}}b^{\frac{1}{6}}x^{\frac{1}{2}}$.

(**16.**) $11x^{-m}y^{-1} \div 2x^{-m}y = \frac{11}{2y^2}$.

(**17.**) $m^{-2}n^{-3}x^my^{-n} \div m^{-1}n^{-3}x^2y^{-1} = \frac{x^{m-2}y^{1-n}}{m}$.

### (PAGE 68, *112.*)

(**1.**) Divide $35a^2x^2 - 30a^3x^3 + 45ax^5$ by $5ax$.

*Quot.*, $7ax - 6a^2x^2 + 9x^4$.

**(2.)** $(30ab^3 - 6a^2b^2) \div 2ab = 15b^2 - 3ab.$

**(3.)** $(3abc + 12abx - 9a^2b) \div 3ab = c + 4x - 3a.$

**(4.)** $(36a^3b^3 + 120a^3b^2 + 60a^2b^2) \div 12a^2b^2 = 3ab + 10a + 5.$

**(5.)** $(10a^2bc - 15a^2b^2c) \div 5abc = 2a - 3ab.$

**(6.)** $(a^2b + ab^2 - ab) \div ab = a + b - 1.$

**(7.)** $(6a^3x^2 - 30a^4x^2 + 90a^3x^3) \div - 6a^3x^2 = 5a - 15x - 1.$

**(8.)** $(144ax - 288a^2) \div 12a = 12x - 24a.$

**(9.)** $(13a^3b^2c - 26a^2b^2c^2 + 39ab^2c^2) \div 13abc = a^2b - 2abc + 3bc.$

**(10.)** $(8a^2x - 4ax^2) \div 8a^2x^2 = x^{-1} - 2^{-1}a^{-1}.$

**(11.)** $(15a^{\frac{1}{2}}b^{\frac{1}{3}} - 30a^{\frac{2}{3}}b^{\frac{2}{5}}x) \div 5a^{\frac{1}{2}}b = 3b^{-\frac{2}{3}} - 6a^{\frac{1}{6}}b^{-\frac{3}{5}}x.$

**(12.)** $(4a^{\frac{1}{5}}x^{\frac{1}{3}} + 2ax - 10a^2x^{\frac{2}{3}}) \div 2ax = 2a^{-\frac{4}{5}}x^{-\frac{2}{3}} + 1 - 5ax^{-\frac{1}{3}}.$

**(13.)** Divide $a^mb^3 + a^{m+1}b^2 + a^{n-2}b$ by $ab$.

*Quot.*, $a^{m-1}b^2 + a^mb + a^{n-3}.$

**(14.)** Divide $12a^{-2} - 8a^2b + 16a^3x - 10a^{-2}y$ by $2a^2$.

*Quot.*, $6a^{-4} - 4b + 8ax - 5a^{-4}y.$

**(15.)** Divide $ax^n + ax^{-n} + ax^{n+2} + ax^{n+m}$ by $x^n$.

*Quot.*, $a + ax^{-2n} + ax^2 + ax^m.$

**(16.)** $(4m^{\frac{a}{b}} - 3m^{-2} + 5m^{\frac{1}{n}}b) \div 7m^{-3} = \frac{4}{7}m^{\frac{a+3b}{b}} - \frac{3}{7}m + \frac{5}{7}m^{\frac{1+3n}{n}}b.$

## (PAGE 70, *114.*)

See pages 57 to 59 in Multiplication, and assign the product and either factor.

**(28.)** Divide $\frac{3}{4}x^5 - 4x^4 + \frac{77}{8}x^3 - \frac{43}{4}x^2 - \frac{33}{4}x + 27$ by $\frac{1}{2}x^2 - x + 3$.

*Quot.*, $\frac{3}{2}x^3 - 5x^2 + \frac{1}{4}x + 9.$

**(29.)** Divide $x^4 - (a - b)x^3 + (p - ab + 3)x^2 + (bp - 3a)x + 3p$ by $x^2 - ax + p$. *Quot.*, $x^2 + bx + 3.$

**(30.)** Divide $ax^3 - (a^2 + b)x^2 + b^2$ by $ax - b$.

*Quot.*, $x^2 - ax - b.$

**(31.)** Divide $x^4 + y^4$ by $x + y$.

*Quot.*, $x^3 - x^2y + xy^2 - y^3 + \frac{2y^4}{x + y}.$

**(32.)** Divide $x^{m+1} + x^my + xy^m + y^{m+1}$ by $x^m + y^m$. *Quot.*, $x + y.$

(**33.**) Divide $x^{4n} + x^{2n}y^{2n} + y^{4n}$ by $x^{2n} + x^n y^n + y^{2n}$.
*Quot.*, $x^{2n} - x^n y^n + y^{2n}$.

(**34.**) $\left(x^5 - \frac{1}{x^5}\right) \div \left(x - \frac{1}{x}\right) = x^4 + x^2 + 1 + \frac{1}{x^2} + \frac{1}{x^4}$.

(**35.**) $\left(\frac{8a^3}{27b^3} - \frac{4a^2x}{3b^2y} + \frac{2ax^2}{by^2} - \frac{x^3}{y^3}\right) \div \left(\frac{2a}{3b} - \frac{x}{y}\right) = \frac{4a^2}{9b^2} - \frac{4ax}{3by} + \frac{x^2}{y^2}$.

(**36.**) $\left(\frac{1}{x^5} - \frac{5}{x^4} + \frac{10}{x^3} - \frac{10}{x^2} + \frac{5}{x} - 1\right) \div \left(\frac{1}{x^2} - \frac{2}{x} + 1\right) = \frac{1}{x^3} - \frac{3}{x^2} + \frac{3}{x} - 1$.

(**37.**) $(1 + a) \div (1 - a) = 1 + 2a + 2a^2 + 2a^3 +$ etc.

(**38.**) $a \div (1 + x) = a - ax + ax^2 - ax^3 +$ etc.

(**39.**) $(a^{m+n} + a^{n+1}b^{m-1} - a^{m-1}b^{n+1} - b^{m+n}) \div (a^{n+1} - b^{n+1}) = a^{m-1} + b^{m-1}$.

## FACTORING.

### (Page 78, *121.*)

Examples under this article can be taken from Art. 111 in Division, either in this collection or in the text-book, or from pages 55 and 56 of the text-book.

### (Page 79, *122.*)

(**1.**) $5x - 5y = 5(x - y)$.

(**2**) $my + ny - cy = (m + n - c)y$.

(**3.**) $7 + 14y = 7(1 + 2y)$.

(**4.**) $12ax^3 - 24mx^2 = 12x^2(ax - 2m)$.

Art. 112 will furnish further exercises, by assigning simply the dividend.

### (Pages 79—81, *123, 124.*)

(**1.**) $m^2 - 2mn + n^2 = (m - n)(m - n)$.

(**2.**) $4a^2 + 12ab + 9b^2 = (2a + 3b)(2a + 3b)$.

See articles 94, 95, and 96, of this collection and also of the text-book, assigning the products as examples under these articles. In giving them the order of the terms may be changed.

## (PAGE 82, *125.*)

Any example in Division may be so stated as to afford an example under this article; or any of the examples in the immediately preceding articles may be so used.

## (PAGES 83, 84, *126, 127.*)

(**1.**) $(m^3 - n^3) \div (m - n) = m^2 + mn + n^2$.

(**2.**) $(x^5 + y^5) \div (x + y) = x^4 - x^3y + x^2y^2 - xy^3 + y^4$.

(**3.**) $(a^5 - x^5y^5) \div (a - xy) = a^4 + a^3xy + a^2x^2y^2 + ax^3y^3 + x^4y^4$.

**4.**) $\left(\frac{a^4}{b^4} - \frac{x^4}{y^4}\right) \div \left(\frac{a}{b} + \frac{x}{y}\right) = \frac{a^3}{b^3} - \frac{a^2x}{b^2y} + \frac{ax^2}{by^2} - \frac{x^3}{y^3}$.

(**5.**) $\left(\frac{a^4}{b^4} - \frac{x^4}{y^4}\right) \div \left(\frac{a}{b} - \frac{x}{y}\right) = \frac{a^3}{b^3} + \frac{a^2x}{b^2y} + \frac{ax^2}{by^2} + \frac{x^3}{y^3}$.

(**6.**) $(9x^2 - 4y^2) \div (3x - 2y) = 3x + 2y$.

(**7.**) $(4x^2 - 4y^2) \div (2x + 2y) = 2x - 2y$.

(**8.**) $(x^{-2} - y^{-2}) \div \left(\frac{1}{x} - \frac{1}{y}\right) = \left(\frac{1}{x^2} - \frac{1}{y^2}\right) \div \left(\frac{1}{x} - \frac{1}{y}\right) = \frac{1}{x} + \frac{1}{y} = x^{-1} + y^{-1}$.

(**9.**) $(a^6 - x^6) \div (a^3 + x^3) = a^3 - x^3$.

(**10.**) $(a^5 + x^{15}) \div (a + x^3) = a^4 - a^3x^3 + a^2x^6 - ax^9 + x^{12}$.

(**11.**) $(27x^{12} - 8y^6) \div (3x^4 - 2y^2) = 9x^8 + 6x^4y^2 + 4y^4$.

(**12.**) $(x^4 - 1) \div (x - 1) = x^3 + x^2 + x + 1$.

(**13.**) $(1 - m^8) \div (1 + m) = 1 - m + m^2 - m^3 + m^4 - m^5 + m^6 - m^7$.

(**14.**) $(m^3y^3 + n^6x^6) \div (my + n^2y^2) = m^2y^2 - mn^2yx^2 + n^4x^4$.

(**15.**) $(81y^{16} - 625x^8) \div (3y^4 \pm 5x^2) = [(3y^4)^4 - (5x^2)^4] \div [(3y^4) \pm (5x^2)] = (3y^4)^3 \mp (3y^4)^2(5x^2) + (3y^4)(5x^2)^2 \mp (5x^2)^3 = 27y^{12} \mp 45y^8x^2 + 75y^4x^4 \mp 125x^6$.

(**16.**) Is $m^9 + x^9$ divisible by $m + x$, or by $m - x$?

*Ans.*, By $m + x$ only.

(**17.**) Is $x^8 - y^8$ divisible by $x + y$, or by $x - y$?

*Ans.*, By both.

(**18.**) $(a^{2x} - y^{2x}) \div (a \mp y)$, ($x$ being a positive integer) $= a^{2x-1} \mp a^{2x-2}y + a^{2x-3}y^2 \pm a^{2x-4}y^3 +$ etc., till the exponent of $a = 0$.

(PAGE 85, ***128.***)

**(19.)** $(x^{-3} - y^{-3}) \div (x^{-1} - y^{-1}) = x^{-2} + x^{-1}y^{-1} + y^{-2}$.

**(20.)** $(x^{-\frac{6}{5}} - m^{-6}) \div (x^{-\frac{1}{5}} + m^{-1}) = x^{-1} - x^{-\frac{4}{5}}m^{-1} + x^{-\frac{3}{5}}m^{-2} -$
$x^{-\frac{2}{5}}m^{-3} + x^{-\frac{1}{5}}m^{-4} - m^{-5}$.

**(21.)** $\left(\frac{1}{x^6} + y^3\right) \div (x^{-2} + y) = \frac{1}{x^4} - \frac{y}{x^2} + y^2$.

**(22.)** Is $x^{-\frac{4}{5}} - y^4$ divisible by $x^{-\frac{1}{5}} \pm y$? *Ans.*, By either.

**(23.)** Is $x^4 + y^4$ divisible by $\sqrt{x} \pm \sqrt{y}$? *Ans.*, By neither.

**(24.)** Is $x^5 + y^5$ divisible by $\sqrt{x} \pm \sqrt{y}$?
*Ans.*, By neither, since $x^5$ is the 10th (an even) power of $\sqrt{x}$.

**(25.)** Is $x^{\frac{5}{2}} + y^{\frac{5}{2}}$ divisible by $\sqrt{x} + \sqrt{y}$? *Ans.*, Yes.

(PAGE 86, ***129.***)

**(1.)** $y^2 - 6y + 5 = (y - 5)(y - 1)$.

**(2.)** $y^2 - 4y - 21 = (y + 3)(y - 7)$.

$y^2 + 12y + 11 = (y + 11)(y + 1)$.

$y^2 + 2y - 48 = (y - 6)(y + 8)$.

$y^2 - 6y + 8 = (y - 4)(y - 2)$.

$y^2 - 7y - 30 = (y - 10)(y + 3)$

(PAGE 87, ***130.***)

**(1--6.)** Use the examples in the preceding article.

**(7.)** $6x^2 - 25x + 14 = 6x^2 - 4x - (21x - 14) = 2x(3x - 2) - 7(3x - 2) = (2x - 7)(3x - 2)$.

**(8.)** $10a^2 + 14ac - 12c^2 = 10a^2 - 6ac + 20ac - 12c^2 = 2a(5a - 3c) + 4c(5a - 3c) = (2a + 4c)(5a - 3c)$.

---

## GREATEST OR HIGHEST C. D.

(PAGE 89, ***132.***)

**(1.)** The G. C. D. of 18, 48, 72, and 66 is 6.

**(2.)** The G. C. D. of 12, 36, 60, and 132 is 12.

**(3.)** The G. C. D. of 72, 108, and 252 is 36.

(**4.**) The G. C. D. of $9abc^3$, and $12bc^4x$ is $3bc^3$.

(**5.**) The G. C. D. of $4a^3b^2x^5y^3$, and $8a^5x^2y^2$ is $4a^3x^2y^2$.

(**6.**) The G. C. D. of $3a^4y^3$, $6a^5x^3y^5$, and $9a^6y^4z$ is $3a^4y^3$.

(**7.**) The G. C. D. of $8ax^2y^4z^5$, $12x^5z^6$, and $24a^3x^3z^2$ is $4x^2z^2$.

(**8.**) The G. C. D. of $x^2 + 2x - 3$, and $x^2 + 5x + 6$ is $x + 3$.

(**9.**) The G. C. D. of $6a^2+11ax+3x^2$, and $6a^2+7ax-3x^2$ is $2a+3x$.

(**10.**) The G. C. D. of $a^4 - x^4$, and $a^3 + a^2x - ax^2 - x^3$ is $a^2 - x^2$.

## (Page 97, *137.*)

(**1.**) Find the H. C. D. of $a^4 + a^3b - ab^3 - b^4$, and $a^4 + a^2b^2 + b^4$.
*H. C. D.*, $a^2 + ab + b^2$.

(**2.**) Find the H. C. D. of $a^2 - 2ax + x^2$, and $a^3 - a^2x - ax^2 + x^3$.
*H. C. D.*, $a^2 - 2ax + x^2$.

(**3.**) Find the H. C. D. of $6x^3 - 8yx^2 + 2y^2x$, and $12x^2 - 15yx + 3y^2$.
*H. C. D.*, $x - y$.

(**4.**) Find the H. C. D. of $36b^2a^6 - 18b^2a^5 - 27b^2a^4 + 9b^2a^3$, and $27b^2a^5 - 18b^2a^4 - 9b^2a^3$. *H. C. D.*, $9b^2a^4 - 9b^2a^3$.

(**5.**) Find the H. C. D. of $36a^2x^5 - 120a^2x^3 + 180a^2x + 96a^2$, and $6a^2x^4 - 12a^2x^2 + 6a^2$. *H. C. D.*, $6(a^2x^2 + 2a^2x + a^2)$.

(**6.**) Find the H. C. D. of $a^4 - 4a^3b + 3a^2b^2$, and $a^2b - 5ab^2 + 4b^3$.
*H. C. D.*, $a - b$.

## (Page 97, *138.*)

(**1.**) Find the H. C. D. of $a^4 - b^4$, $a^3 - a^2b + 3a - 3b$, and $a^2 + a - ab - b$. *H. C. D.*, $a - b$.

(**2.**) Find the H. C. D. of $6x^4 - 6$, $3x^5y + 3x^3y$, $12ax^8 - 12a$, and $3x^5 + 6x^3 + 3x$. *H. C. D.*, $3x^2 + 3$.

---

## LEAST OR LOWEST COMMON MULTIPLE.

## (Page 99, *140.*)

(**1.**) Find the L. C. M. of $8a$, $4a^2$, and $12ab$. *L. C. M.*, $24a^2b$.

(**2.**) Find the L. C. M. of $a^2 - b^2$, $a + b$, and $a^2 + b^2$.
*L. C. M.*, $a^4 - b^4$.

(**3.**) Find the L. C. M. of $27a$, $15b$, $9ab$, and $3a^2$. *L. C. M.*, $135a^2b$.

(**4.**) Find the L. C. M. of $a^3 + 3a^2b + 3ab^2 + b^3$, $a^2 + 2ab + b^2$, $a^2 - b^2$. *L. C. M.*, $a^4 + 2a^3b - 2ab^3 - b^4$.

(**5.**) Find the L. C. M. of $a+b$, $a - b$, $a^2 + ab + b^2$, and $a^2 - ab + b^2$. *L. C. M.*, $a^6 - b^6$.

(**6.**) Find the L. C. M. of $12a^2y(a + b)$, $6a^3y^2 + 12a^2by^2 + 6ab^2y^2$, and $4a^2y^2$. *L. C. M.*, $12a^2y^2(a + b)^2$.

(**7.**) Find the L. C. M. of $x^3 - x^2y - xy^2 + y^3$, $x^3 - x^2y + xy^2 - y^3$, and $x^4 - y^4$. *L. C. M.*, $x^5 - xy^4 - x^4y + y^5$.

(**8.**) Find the L. C. M. of $(a + b)^2$, $(a^2 - b^2)$, $(a - b)^2$, and $a^3 + 3a^2b + 3ab^2 + b^3$. *L. C. M.*, $(a + b)(a^2 - b^2)^2$.

(**9.**) Find the L. C. M. of $x^4 - 5x^3 + 9x^2 - 7x + 2$, and $x^4 - 6x^2 + 8x - 3$. *L. C. M.* , $x^5 - 2x^4 - 6x^3 + 20x^2 - 19x + 6$.

(**10.**) Find the L. C. M. of $a^3 + 2a^2b - ab^2 - 2b^3$, $a^3 - 2a^2b - ab^2 + 2b^3$, and $3a^3 - a^2 - 3ab^2 + b^2$.
*L. C. M.*, $(a^2 - 4b^2)(3a - 1)(a^2 - b^2)$.

(**11.**) Find the L. C. M. of $4(1 - x^2)$, $8(1 - x)$, $8(1 + x)$, and $4(1 + x^2)$. *L. C. M.*, $8(1 - x^4)$.

(**12.**) Find the L. C. M. of $x^2 + 5x + 4$, $x^2 + 2x - 8$, and $x^2 + 7x + 12$. *L. C. M.*, $x^4 + 6x^3 + 3x^2 - 26x - 24$.

---

## FRACTIONS.

(PAGE 106, ***158.***)

(**1.**) What is the essential sign of $\frac{3a^3 - 3ab^2}{5ab + 5b^2}$, when $a = -2$, and $b = 3$? *Ans.*, +.

(**2.**) What is the essential sign of $\frac{y^3 - x^3}{y^2 - 2xy + x^2}$, when $y = -3$, $x = 1$? *Ans.*, —.

(**3.**) What is the essential sign of $\frac{9x^2y + 9xy^2}{3x^2 + 6xy + 3y^2}$, when $y = 2$, and $x = -3$? *Ans.*, +.

(**4.**) What is the essential sign of $\frac{-(2a^2b + 3by)}{-b^2y^2 + ay}$, when $a = -2$, $b = -1$, $y = -3$? *Ans.*, +.

## (Page 106, *159.*)

(1, 2.) Reduce the following fractions to their lowest terms:

$$\frac{4a^3x^2}{6a^4} = \frac{2x^2}{3a}.$$

$$\frac{6a^2x^2}{8ax^3} = \frac{3a}{4x}.$$

$$\frac{6a^4x^2}{8a^2xy^4} = \frac{3a^2x}{4y^4}.$$

$$\frac{9x^4y^3z^5}{12x^3y^4z^5} = \frac{3x}{4y}.$$

$$-\frac{12x^2y^2z^4}{8x^2z^3} = -\frac{3y^2z}{2}.$$

$$\frac{8ab^2}{12ab^2 + 4abc} = \frac{2b}{3b + c}.$$

$$\frac{2a^2cx^2 + 2acx}{10ac^2x} = \frac{ax + 1}{5c}$$

$$\frac{5a^2b + 5ab^2}{5abc + 5abd} = \frac{a + b}{c + d}.$$

(3.)

$$\frac{a - b}{a^3 + b^3} = \frac{1}{a^2 + ab + b^2}.$$

(4.)

$$\frac{x^4 + a^2x^2 + a^4}{x^4 + ax^3 - a^3x - a^4} = \frac{x^2 - ax + a^2}{x^2 - a^2}.$$

(5.) $$\frac{7a^2 - 23ab + 6b^2}{5a^3 - 18a^2b + 11ab^2 - 6b^3} = \frac{7a - 2b}{5a^2 - 3ab + 2b^2}.$$

(6.)

$$\frac{a^4 - b^4}{a^6 - b^6} = \frac{a^2 + b^2}{a^4 + a^2b^2 + b^4}.$$

(7.)

$$\frac{y^4 - x^4}{y^3 - y^2x - yx^2 + x^3} = \frac{y^2 + x^2}{y - x}.$$

(8.) $$\frac{a^3 - 3a^2x + 3ax^2 - x^3}{a^2 - x^2} = \frac{a^2 - 2ax + x^2}{a + x}.$$

(9.) $$\frac{a^3 + 2ba^2 + 3b^2a^2}{2a^4 - 3ba^3 - 5b^2a^2} = \frac{a + 2b + 3b^2}{2a^2 - 3ba - 5b^2}.$$

(10.)

$$\frac{a^2 - 2ax + x^2}{a^3 - a^2x - ax^2 + x^3} = \frac{1}{a + x}.$$

(11.)

$$\frac{a^3 - b^2a}{a^2 + 2ab + b^2} = \frac{a^2 - ab}{a + b}$$

(12.) $$\frac{20x^4 + x^2 - 1}{25x^4 + 5x^3 - x - 10x^2 + 1} = \frac{4x^2 + 1}{5x^2 + x - 1}.$$

Examples like 13—16 may be made from the preceding, by asking what factor reduces the result to the given form.

## (Page 109, *160.*)

(1.) $$\frac{3a^2b^2 + 6ab - 2x + 2c}{3ab} = ab + 2 - \frac{2x - 2c}{3ab}.$$

(2.) $$\frac{21ax^2 - 4x + 9}{7a} = 3x^2 - \frac{4x - 9}{7a}.$$

(3.)

$$\frac{8x^2y^2 - 3ax - 6b}{4x^2} = 2y^2 - \frac{3ax + 6b}{4x^2}.$$

(4.)

$$\frac{x^4 - a^4}{x^2 + a^2} = x^2 - a^2.$$

(5.) $$\frac{27a^3 + 3b^2 - 4x - 9a^2}{9a^2} = 3a - 1 + \frac{3b^2 - 4x}{9a^2}.$$

(6.)

$$\frac{x^4 - 3x^2y^2 + 4ax}{x^2 - 3y^2} = x^2 + \frac{4ax}{x^2 - 3y^2}.$$

(7.)

$$\frac{x^6 + 3ax^2 - a^6 - b}{x^3 + a^3} = x^3 - a^3 + \frac{3ax^2 - b}{x^3 + a^3}.$$

(8.) $$\frac{3x^2 - 12ax + y - 9x}{3x} = x - 4a - 3 + \frac{y}{3x}.$$

(9.)

$$\frac{x^8 + x^4y^4 + y^8}{x^4 + x^2y^2 + y^4} = x^4 - x^2y^2 + y^4.$$

(10.)

$$\frac{x^2 - a^2 + b}{x + a} = x - a + \frac{b}{x + a}.$$

(11.) $$\frac{x^3 + 6x^2 + 12x + 10}{x + 2} = x^2 + 4x + 4 + \frac{2}{x + 2}.$$

(12.) $$\frac{a^5 + b^5}{a + b} = a^4 - a^3b + a^2b^2 - ab^3 + b^4.$$

## (PAGE 110, *161.*)

(5.) Express $\frac{x - y}{a(x - y)^{-1}}$ in the integral form.

*Result,* $a^{-1}(x - y)^2$.

(6.) $\frac{5a^2m^3}{6xy^2} = 5\cdot 6^{-1}a^2m^3x^{-1}y^{-2}$. $\frac{x - y}{(x - y)^3} = (x - y)^{-2}$. $\frac{a^2b^3}{a^{-3}b^2x^{-1}} = a^5bx$. $\frac{x - y}{a^{-2}(x^2 - y^2)^{-2}} = a^2(x - y)^3(x + y)^2$.

## (PAGE 111, *162.*)

The examples in 160 above or in the text-book will afford examples under this article by assigning the results to be reduced back to the forms in the examples.

## (PAGE 112, *163.*)

(1.) $\frac{2}{3}$, $\frac{3a}{4}$, and $\frac{x - y}{b} = \frac{8b}{12b}$, $\frac{9ab}{12b}$, and $\frac{12x - 12y}{12b}$

(2.) $\frac{2x}{3y}$, $\frac{3x}{5z}$, and $a = \frac{10xz}{15yz}$, $\frac{9xy}{15yz}$, and $\frac{15ayz}{15yz}$.

(3.) $\frac{a}{x}$, $\frac{x}{y}$, and $\frac{y}{z} = \frac{ayz}{xyz}$, $\frac{x^2z}{xyz}$, and $\frac{xy^2}{xyz}$.

(4.) $\frac{1}{2}, \frac{x^2}{3}$, and $\frac{x^2+z^2}{x+z} = \frac{3x+3z}{6x+6z}, \frac{2x^3+2x^2z}{6x+6z}$, and $\frac{6x^2+6z^2}{6x+6z}$.

(5.) $\frac{x+y}{x-y}$, and $\frac{x-y}{x+y} = \frac{x^2+2xy+y^2}{x^2-y^2}$, and $\frac{x^2-2xy+y^2}{x^2-y^2}$.

(6.) $\frac{x}{3}, \frac{x+1}{4}$, and $\frac{x-1}{1+x} = \frac{4x^2+4x}{12x+12}, \frac{3x^2+6x+3}{12x+12}$, and $\frac{12x-12}{12x+12}$

## (Page 114, *164.*)

(5.) $\frac{m}{ac}, \frac{n}{b^2c}$, and $\frac{r}{c^2d} = \frac{b^2cdm}{ab^2c^2d}, \frac{acdn}{ab^2c^2d}$, and $\frac{ab^2r}{ab^2c^2d}$.

(6.) $\frac{x+y}{x-y}, \frac{x-y}{x+y}$, and $\frac{x^2+y^2}{x^2-y^2} = \frac{(x+y)^2}{x^2-y^2}, \frac{(x-y)^2}{x^2-y^2}$, and $\frac{x^2+y^2}{x^2-y^2}$.

(7.) $\frac{4a}{a+x}, \frac{3b}{x^2}, \frac{2ax}{a-x} = \frac{4a^2x^2-4ax^3}{a^2x^2-x^4}, \frac{3a^2b-3bx^2}{a^2x^2-x^4}, \frac{2a^2x^3+2ax^4}{a^2x^2-x^4}$.

(8.) $\frac{a}{a^2-x^2}, \frac{3b}{4a-4x}, \frac{5x}{a+x} = \frac{4a}{4a^2-4x^2}, \frac{3ab+3bx}{4a^2-4x^2}, \frac{20ax-20x^2}{4a^2-4x^2}$.

(9.) $\frac{1}{a^2+2ax+x^2}, \frac{1}{a^2-x^2}$, and $\frac{5y}{a^4-x^4} = \frac{a^3+ax^2-a^2x-x^3}{a^5-ax^4+a^4x-x^5}$,
$\frac{a^3+ax^2+a^2x+x^3}{a^5-ax^4+a^4x-x^5}$, and $\frac{5ay+5xy}{a^5-ax^4+a^4x-x^5}$.

(10.) $\frac{x}{a+b}, \frac{y}{a-b}$, and $\frac{z}{a^2-b^2} = \frac{x(a-b)}{a^2-b^2}, \frac{y(a+b)}{a^2-b^2}$, and $\frac{z}{a^2-b^2}$.

(11.) $\frac{m-n}{m+n}, \frac{m+n}{m-n}, \frac{m^2n^2}{m^2-n^2} = \frac{(m-n)^2}{m^2-n^2}, \frac{(m+n)^2}{m^2-n^2}$, and $\frac{m^2n^2}{m^2-n^2}$.

(12.) $\frac{a^2}{a+b}, \frac{a+b}{a-b}$, and $\frac{a^2b^2}{a^2-b^2} = \frac{a^2(a-b)}{a^2-b^2}, \frac{(a+b)^2}{a^2-b^2}$, and $\frac{a^2b^2}{a^2-b^2}$.

## (Page 115, *165.*)

(1.)

$$\frac{\frac{a}{b}}{m+n} = \frac{a}{bm+bn}.$$

(2.)

$$\frac{\frac{a}{b}+\frac{c}{d}}{\frac{e}{f}-\frac{g}{h}} = \frac{fh(ad+bc)}{bd(eh-fg)}.$$

(3.)

$$\frac{\frac{3x}{2x-2}}{\frac{2x}{x-1}} = \frac{3}{4}.$$

(4.)

$$\frac{1}{x+\frac{1}{1+\frac{x+1}{3-x}}} = \frac{4}{3(x+1)}.$$

(5.)

$$\frac{y - x + \frac{a}{2}}{7\frac{3}{4}} = \frac{4y - 4x + 2a}{31}.$$

(6.)

$$\frac{\frac{a+1}{a-1} + \frac{a-1}{a+1}}{\frac{a+1}{a-1} - \frac{a-1}{a+1}} = \frac{a^2+1}{2a}.$$

(7.)

$$\frac{a + b + \frac{b^2}{a}}{a + b + \frac{a^2}{b}} = \frac{b}{a}.$$

(8.)

$$\frac{\frac{1}{a} + \frac{1}{ab^3}}{b - 1 + \frac{1}{b}} = \frac{b+1}{ab^2}.$$

---

## ADDITION.

### (Page 117, *166.*)

(1.)

$$\frac{a}{2} + \frac{a}{3} + \frac{a}{6} = a.$$

(2.)

$$\frac{x+y}{2} + \frac{x-y}{2} = x.$$

(3.)

$$\frac{x}{x+y} + \frac{y}{x-y} = \frac{x^2+y^2}{x^2-y^2}.$$

(4.)

$$\frac{a-b}{ab} + \frac{b-c}{bc} + \frac{c-a}{ac} = 0.$$

(5.)

$$\frac{x}{x-3} + \frac{x}{x+3} = \frac{2x^2}{x^2-9}.$$

(6.)

$$\frac{a+b}{a-b} + \frac{a-b}{a+b} = \frac{2a^2+2b^2}{a^2-b^2}.$$

(7.) $\frac{a^2 - ab + b^2}{a-b} + \frac{a^2 + ab + b^2}{a+b} = \frac{2a^3}{a^2-b^2}.$

(8.) $\frac{2}{x^3 + x^2 + x + 1} + \frac{3}{x^3 - x^2 + x - 1} = \frac{5x+1}{x^4-1}.$

(9.) $\frac{4x^2 - 11x - 5}{5x+5} + \frac{3x+1}{x+1} = \frac{4}{5}x.$

(10.) $\frac{1}{x+3} + \frac{x+1}{x^2 - 3x + 9} + \frac{x^2+x+1}{x^3+27} = \frac{3x^2+2x+13}{x^3+27}.$

### (Page 119, *167.*)

(1.) Add $2x + \frac{x-2}{3}$ and $3x + \frac{2x-3}{4}$. *Sum*, $5x + \frac{10x-17}{12}$.

(2.) Add $4x$, $\frac{7x}{9}$ and $2 + \frac{x}{5}$. *Sum*, $4x + 2 + \frac{44x}{45}$.

(3.) Add $5x - \frac{2x}{7}$ and $\frac{5x}{9} - 4x$. *Sum*, $x + \frac{17x}{63}$.

(4.) Add $2x$, $3x + \frac{3z}{5}$, and $x + \frac{2z}{9}$. *Sum*, $6x + \frac{37z}{45}$.

(5.) Add $3 + \frac{2a}{5}$, $5 - \frac{3a - 2x}{x}$, and $7 + \frac{x - a}{a}$.

*Sum*, $16 + \frac{2a^2x - 15a^2 + 5x^2}{5ax}$.

(6.) Add $5x + \frac{x - 2}{3}$, and $4x - \frac{2x - 3}{5x}$.

*Sum*, $9x + \frac{5x^2 - 16x + 9}{15x}$.

(7.) $\frac{2}{x + a} + \frac{3a}{(x + a)^2} + \frac{3a - 2x}{x^2 - 2ax + 3a^2} = \frac{18a^3}{x^4 + 4a^3x + 3a^4}$.

(8.) $\frac{1}{4a^3(a + x)} + \frac{1}{4a^3(a - x)} + \frac{1}{2a^2(a^2 + x^2)} = \frac{1}{a^4 - x^4}$.

(9.) $\frac{1}{a(a - b)(a - c)} + \frac{1}{b(b - a)(b - c)} + \frac{1}{c(c - a)(c - b)} = \frac{1}{abc}$.

(10.) $\frac{x + 2}{x^2 + x - 2} + \frac{x + 3}{x^2 + 4x + 3} + \frac{2x + 8}{x^3 + 4x^2 - x - 4} = \frac{2}{x - 1}$.

## SUBTRACTION.

### (Page 121, *168*.)

(1.) From $\frac{a + x}{2}$ take $\frac{a - x}{2}$. *Rem.*, $x$.

(2.) From $\frac{7x^2 - 9x + 11}{10}$ take $\frac{9x^2 - 11x + 15}{20}$.

*Rem.*, $\frac{5x^2 - 7x + 7}{20}$.

(3.) From $\frac{4x}{5}$ take $\frac{3x + 1}{x + 1}$. *Rem.*, $\frac{4x^2 - 11x - 5}{5x + 5}$.

(4.) From $\frac{ax}{a^2 - x^2}$ take $\frac{a - x}{a + x}$. *Rem.*, $\frac{3ax - a^2 - x^2}{a^2 - x^2}$.

(5.) From $\frac{13a - 5b}{4} - \frac{3a}{5}$ take $\frac{7a - 2b}{6}$. *Rem.*, $\frac{89a - 55b}{60}$.

(6.) From $\frac{4a}{13b}+\frac{6b}{13a}$ take $\frac{7a}{2b}$. *Rem.*, $\frac{12b^2-83a^2}{26ab}$.

(7.) From $\frac{a+b}{a-b}$ subtract $\frac{a-b}{a+b}$. *Rem.*, $\frac{4ab}{a^2-b^2}$.

(8.) From $\frac{1}{a-x}$ subtract $\frac{1}{a+x}$. *Rem.*, $\frac{2x}{a^2-x^2}$.

(9.) From $\frac{4x+2}{3}$ subtract $\frac{2x-3}{3x}$. *Rem.*, $\frac{4x^2+3}{3x}$.

(10.) $\frac{2}{x}-\frac{3}{2x-1}-\frac{2x-3}{4x^2-1}=\frac{-2}{(4x^2-1)x}$.

(11.) $\frac{1}{x-1}-\frac{1}{x+2}-\frac{3}{(x+2)^2}=\frac{9}{(x-1)(x+2)^2}$.

(12.) $\frac{b-a}{x-b}-\frac{a-2b}{x+b}+\frac{3x(a-b)}{x^2-b^2}=\frac{ax-b^2}{x^2-b^2}$.

(13.) $\frac{3+2x}{2-x}-\frac{2-3x}{2+x}+\frac{16x-x^2}{x^2-4}=\frac{1}{x+2}$.

(14.) $\frac{a}{a-x}+\frac{3a}{a+x}-\frac{2ax}{a^2-x^2}=\frac{4a}{a+x}$.

(15.) $\frac{a+b}{(b-c)(c-a)}+\frac{b+c}{(c-a)(a-b)}+\frac{c+a}{(a-b)(b-c)}=0.$

(16.) $\frac{a-b}{a+b}+\frac{b-c}{b+c}+\frac{c-a}{c+a}+\frac{(a-b)(b-c)(c-a)}{(a+b)(b+c)(c+a)}=0.$

(17.) From $2x+\frac{5x-2}{7}$ subtract $3x-\frac{4x+5}{6}$.

*Rem.*, $\frac{16x+23}{42}$.

(18.) Subtract $4x-\frac{2x-3}{5}$ from $5x+\frac{x-2}{3}$.

*Rem.*, $x+\frac{11x-19}{15}$.

(19.) Subtract $\frac{a+x}{a(a-x)}$ from $a+\frac{a-x}{a(a+x)}$.

*Rem.*, $a-\frac{4x}{a^2-x^2}$.

(20.) From $3x$ take $\frac{3a+12x}{5}$. *Rem.*, $\frac{3x-3a}{5}$.

## MULTIPLICATION.

### (Page 124, *170*.)

(1.)

$$\frac{4c}{2a+c} \times (a-2b) = \frac{4ac-8bc}{2a+c}.$$

(2.)

$$\frac{2a+b}{a^2-b^2} \times (a-b) = \frac{2a+b}{a+b}.$$

(3.)

$$\frac{abc}{3(x^4-y^4)} \times (x^2+y^2) = \frac{abc}{3(x^2-y^2)}.$$

(4.)

$$\frac{b+c}{b-c} \times (a+c) = \frac{ab+ac+bc+c^2}{b-c}.$$

(5.)

$$\frac{x+y}{c} \times (x-y) = \frac{x^2-y^2}{c}.$$

(6.)

$$\frac{x^2+xy+y^2}{x+y} \times (x-y) = \frac{x^3-y^3}{x+y}.$$

(7.)

$$\frac{a}{a^3-x^3} \times (a-x)^2 = \frac{a^2-ax}{a^2+ax+x^2}.$$

(8.)

$$\frac{3xy}{5x^2-5y^2} \times 5x(x-y) = \frac{3x^2y}{x+y}.$$

(9.)

$$\frac{1-x-x^2}{c-d} \times (c-d) = 1-x-x^2.$$

(10.)

$$\frac{2y}{a^2-y^2} \times 5(a^2-y^2) = 10y.$$

(11.)

$$\frac{m^3-n^3}{2x-3y} \times (2x-3y) = m^3-n^3.$$

(12.)

$$\frac{5a}{1-y^2} \times (1-y^4) = 5a(1+y^2).$$

(13.)

$$\frac{a-b}{a+b} \times (a+b)^2 = a^2-b^2.$$

### (Page 126, *171*.)

(1.)

$$\frac{(m+x)^3}{a^3+b^3} \times \frac{a+b}{(m+x)^2} = \frac{m+x}{a^2-ab+b^2}.$$

(2.)

$$\frac{2x}{3cm} \times \frac{5m}{4cx} = \frac{5}{6c^2}.$$

(3.)

$$\frac{3x^2}{10y} + \frac{5y}{9x} = \frac{x}{6}.$$

(4.)

$$\frac{5x}{8y} \times \frac{3x}{2y^2} = \frac{15x^2}{16y^3}.$$

(5.)

$$\frac{3c^2m}{4ax^2} \times \frac{5cm^2}{2x} = \frac{15c^3m^3}{8ax^3}.$$

(6.)

$$\frac{a^2-x^2}{3a} \times \frac{7x^2}{a-x} = \frac{7ax^2+7x^3}{3a}.$$

(7.)

$$\frac{3x^2-5x}{14} \times \frac{21x}{3a-1} = \frac{9x^3-15x^2}{6a-2}.$$

(8.)

$$\frac{3x^2}{5x-10} \times \frac{15x-30}{2x} = \frac{9x}{2}.$$

(9.)

$$\frac{2a-2x}{3ab}\times\frac{3ax}{5a-5x}=\frac{2x}{5b}.$$

(10.)

$$\frac{a^4-x^4}{a^2-y^2}\times\frac{a+y}{a^2+x^2}\times\frac{a-y}{a-x}=a+x.$$

(11.)

$$\frac{a^2-x^2}{a+b}\times\frac{a^2-b^2}{a+x}\times\frac{a}{ax-x^2}=\frac{a^2-ab}{x}.$$

(12.)

$$\frac{1+a+a^2}{1-b+b^2}\times\frac{1-a}{1+b}=\frac{1-a^3}{1+b^3}.$$

(13.)

$$\frac{4ax}{3by}\times\frac{a^2-x^2}{c^2-x^2}\times\frac{bc+bx}{a^2-ax}=\frac{4x(a+x)}{3y(c-x)}.$$

(14.)

$$\frac{a}{b}\times\frac{a^m}{b^n}=\frac{a^{m+1}}{b^{n+1}}.$$

(15.)

$$\frac{c}{x}\times\frac{x^m}{c^n}=\frac{x^{m-1}}{c^{n-1}}.$$

(16.)

$$\frac{x^{\frac{2}{3}}-y^{\frac{2}{3}}}{a^3-b^3}\times\frac{a^2+ab+b^2}{x^{\frac{1}{3}}+y^{\frac{1}{3}}}=\frac{x^{\frac{1}{3}}-y^{\frac{1}{3}}}{a-b}.$$

(17.)

$$\frac{3a^{-1}x^3}{10a^mxy^3}\times\frac{5a^{m+1}y^{-1}}{6x^{-2}y^{-3}}=\frac{x^4}{4y}.$$

(18.)

$$\frac{-4ax}{cy}\times\frac{cxy-y^2}{6x^2+6xy}=\frac{2a(y-cx)}{3c(x+y)}.$$

(19.) $$\frac{a^2+ax+x^2}{a^3-a^2x+ax^2-x^3}\times\frac{a^2-ax+x^2}{a+x}=\frac{a^4+a^2x^2+x^4}{a^4-x^4}.$$

(20.) $$\left(a+\frac{x}{5}\right)\times\left(a-\frac{x}{3}\right)=a^2-\frac{2ax+x^2}{15}.$$

(21.) $$\left(a+\frac{x}{b}\right)\times\left(a-\frac{y}{b}\right)=\frac{a^2b^2+abx-aby-xy}{b^2}=a^2+\frac{ax-ay}{b}-\frac{xy}{b^2}.$$

(22.) $$(x^2+x+1)\times\left(\frac{1}{x^2}-\frac{1}{x}+1\right)=x^2+1+\frac{1}{x^2}.$$

(23.) $$\left(x+1+\frac{1}{x}\right)\times\left(x-1+\frac{1}{x}\right)=x^2+1+\frac{1}{x^2}.$$

(24.) $$\left(\frac{4a}{3x}+\frac{3x}{2b}\right)\times\left(\frac{2b}{3x}+\frac{3x}{4a}\right)=\frac{8ab}{9x^2}+2+\frac{9x^2}{8ab}.$$

---

## DIVISION.

(Pages 130—136, ***173—175.***)

The examples in Multiplication, either in the text-book or as given above, will afford sufficient exercise in this subject.

## INVOLUTION.

### (PAGE 141, *192.*)

**(1—6.)** $(3x^2)^3 = 27x^6$. $(-2a^{\frac{1}{5}})^5 = -32a$. $(-3xy^3)^2 = 9x^2y^6$. $(2a^{-1}y^{\frac{3}{2}})^2 = 2a^{-2}y^3$. $(-\frac{3}{4}x^{-2})^2 = \frac{9}{16x^4}$. $\left(\frac{2x^m}{3y}\right)^2 = \frac{4x^{2m}}{9y^2}$.

**(7—10.)** $(-2a^2x^m)^3 = -8a^6x^{3m}$. $(11x^{-\frac{2}{3}})^3 = \frac{1331}{x^2}$. $(-\frac{1}{5}x^{2n-1})^3 = -\frac{1}{125}x^{6n-3}$. $\left(-\frac{a^{-1}x^2}{b^{-2}y}\right)^3 = -\frac{b^6x^6}{a^3y^3}$.

**(11.)** $(a + 2b)^2 = a^2 + 4ab + 4b^2$.

**(12.)** $(a^2 - x^2)^3 = a^6 - 3a^4x^2 + 3a^2x^4 - x^6$.

**(13.)** $(3x^2 + 2x + 5)^2 = 9x^4 + 12x^3 + 34x^2 + 20x + 25$.

**(14.)** $(3x - 5)^3 = 27x^3 - 135x^2 + 225x - 125$.

**(15.)** $(a + 3b)^4 = a^4 + 12a^3b + 54a^2b^2 + 108ab^3 + 81b^4$.

**(16.)** $(1 - 2x)^5 = 1 - 10x + 40x^2 - 80x^3 + 80x^4 - 32x^5$.

**(17.)** $(x^2 - 2xy + y^2)^3 = x^6 - 6x^5y + 15x^4y^2 - 20x^3y^3 + 15x^2y^4 - 6xy^5 + y^6$.

**(18.)** $\left(x - \frac{1}{x} - 1\right)^2 = x^2 - 2x + \frac{1}{x^2} + \frac{2}{x} - 1$.

**(19.)** $(\frac{1}{2}a - \frac{2}{3}b)^3 = \frac{1}{8}a^3 - \frac{8}{27}b^3 - \frac{1}{2}a^2b + \frac{2}{3}ab^2$.

**(20.)** $2(a^{\frac{1}{2}} - 3a^{-\frac{1}{2}})^3 = 8a^{\frac{3}{2}} - 36a^{\frac{1}{2}} + 54a^{-\frac{1}{2}} - 27a^{-\frac{3}{2}}$.

**(21.)** $(e^x - e^{-x})^3 = e^{3x} - e^{-3x} - 3(e^x - e^{-x})$.

### (PAGE 143, *193.*)

**(1—4.)** $(3xy^{-2})^3 = 27x^3y^{-6}$. $(2a^{\frac{2}{5}})^n = (2)^n a^{\frac{2n}{5}}$. $(12m^{12}y^{-3})^{\frac{2}{3}} = (12)^{\frac{2}{3}}m^8y^{-2}$. $(2x^ay^b)^{\frac{m}{n}} = (2)^{\frac{m}{n}}x^{\frac{am}{n}}y^{\frac{bm}{n}}$.

**(5.)** $\left(\frac{3}{5}b^3y\right)^2 = \frac{9}{25}b^6y^2$. $\left(-\frac{2xy^2}{3a^2b}\right)^2 = \frac{4x^2y^4}{9a^4b^2}$. $\left(-\frac{3m^{-2}b^{-\frac{1}{3}}}{2x^{-\frac{1}{2}}y^{-1}}\right)^3 = -\frac{27x^{\frac{3}{2}}y^3}{8m^6b}$.

**(6—8.)** $(27a^{10}y^{-12})^{\frac{2}{3}} = \frac{9a^{\frac{20}{3}}}{y^8}$. $(64m^{12}y^6)^{-\frac{5}{6}} = \frac{1}{32m^{10}y^5}$. $(4a^{-\frac{1}{2}}y^{-2})^{-4}$ $= \frac{1}{256}a^2y^8$.

**(9.)** $(125a^3b^2)^{\frac{2}{3}} = 25a^2b^{\frac{4}{3}}$. $(64x^{-4}y^{\frac{4}{3}})^{-\frac{1}{2}} = \frac{x^2}{8y^{\frac{2}{3}}}$. $(5a^{\frac{1}{3}}x^{-\frac{2}{3}})^{-\frac{1}{2}} = \frac{x^{\frac{1}{3}}}{(5)^{\frac{1}{2}}a^{\frac{1}{6}}}$.
$(m^{\frac{1}{n}}n^{\frac{m}{n}}r^{n-1})^{-\frac{a}{b}} = m^{-\frac{a}{bn}}n^{-\frac{am}{bn}}r^{-\frac{a(n-1)}{b}}$

## (PAGE 143, *194.*)

**(1.)** $(a+b)^7 = a^7 + 7a^6b + 21a^5b^2 + 35a^4b^3 + 35a^3b^4 + 21a^2b^5 + 7ab^6 + b^7$.

**(2.)** $(a-b)^4 = a^4 - 4a^3b + 6a^2b^2 - 4ab^3 + b^4$.

**(3.)** $(2a-3x)^3 = 8a^3 - 36a^2x + 54ax^2 - 27x^3$.

**(4.)** $(x+y)^{-3} = x^{-3} - 3x^{-4}y + 6x^{-5}y^2 - 10x^{-6}y^3 +$, etc.

**(5.)** $(a-b)^{\frac{3}{4}} = a^{\frac{3}{4}} - \frac{3}{4}a^{-\frac{1}{4}}b + \frac{3}{32}a^{-\frac{5}{4}}b^2 - \frac{5}{128}a^{-\frac{9}{4}}b^3 +$, etc.

**(6.)** $(1-y)^m = 1 - my + \frac{m(m-1)}{1\cdot 2}y^2 - \frac{m(m-1)(m-2)y^3}{1\cdot 2\cdot 3} + \frac{m(m-1)(m-2)(m-3)}{1\cdot 2\cdot 3\cdot 4}y^4$, etc.

**(7.)** $(x+a)^{\frac{1}{2}} = x^{\frac{1}{2}} + \frac{a}{1\cdot 2\cdot x^{\frac{1}{2}}} - \frac{1\cdot a}{1\cdot 2\cdot 4\cdot x^{\frac{3}{2}}} + \frac{1\cdot 3\cdot a^3}{1\cdot 2\cdot 4\cdot 6x^{\frac{5}{2}}} - \frac{1\cdot 3\cdot 5\cdot a^5}{1\cdot 2\cdot 4\cdot 6\cdot 8\cdot x^{\frac{7}{2}}} +$, etc.

**(8.)** $\frac{m}{\sqrt{b^2+c^4}} = m(b^2+c^4)^{-\frac{1}{2}} =$

$$m\left(\frac{1}{b} - \frac{1\cdot c^4}{1\cdot 2\cdot b^3} + \frac{1\cdot 3\cdot c^8}{1\cdot 2\cdot 4\cdot b^5} - \frac{1\cdot 3\cdot 5\cdot c^{12}}{1\cdot 2\cdot 4\cdot 6\cdot b^7} +, \text{etc.}\right)$$

**(9.)** $(c^2-x^2)^{\frac{3}{4}} = c^{\frac{3}{2}} - \frac{3x^2}{2^2c^{\frac{1}{2}}} - \frac{3x^4}{2^5c^{\frac{5}{2}}} - \frac{5x^6}{2^7c^{\frac{9}{2}}} - \frac{7x^8}{2^9c^{\frac{13}{2}}} -$, etc.

**(10.)** $(a^2-ax)^{-\frac{3}{10}} = \frac{1}{a^{\frac{3}{5}}}\left(1-\frac{x}{a}\right)^{-\frac{3}{10}} = \frac{1}{a^{\frac{3}{5}}}\left\{1 + \frac{3}{10}\cdot\frac{x}{a} + \frac{3\cdot 13}{(10)^2}\cdot\frac{x^2}{1\cdot 2\cdot a^2} + \frac{3\cdot 13\cdot 23}{(10)^3}\cdot\frac{x^3}{1\cdot 2\cdot 3\cdot a^3} + \frac{3\cdot 13\cdot 23\cdot 33}{(10)^4}\cdot\frac{x^4}{1\cdot 2\cdot 3\cdot 4\cdot a^4} +, \text{etc.}\right\}$

**(11.)** $\frac{n}{\sqrt{b^2 - c^2e^2}} = \frac{n}{b}\left(1 - \frac{c^2e^2}{b^2}\right)^{-\frac{1}{2}} = \frac{n}{b}\left\{1 + \frac{1}{2} \cdot \frac{c^2e^2}{b^2} + \frac{1 \cdot 3}{2 \cdot 4} \cdot \frac{c^4e^4}{b^4} + \frac{1 \cdot 3 \cdot 5}{2 \cdot 4 \cdot 6} \cdot \frac{c^6e^6}{b^6} + \text{etc.}\right\}$

**(12.)** $\frac{a^2}{(a + x)^2} = a^2(a + x)^{-2} = 1 - \frac{2x}{a} + \frac{3x^2}{a^2} - \frac{4x^3}{a^3} + \frac{5x^4}{a^4} -$, etc.

**(13.)** $(a + b - c + d)^2 = a^2 + 2ab + b^2 - 2ac - 2bc + c^2 + 2ad + 2bd - 2cd + d^2$.

**(14.)** $(x^2 + 3y^2)^5 = x^{10} + 15x^8y^2 + 90x^6y^4 + 270x^4y^6 + 405x^2y^8 + 243y^{10}$.

(Page 152, ***200.***)

**(15.)** $(x + y)^6 = x^6 + 6x^5y + 15x^4y^2 + 20x^3y^3 + 15x^2y^4 + 6xy^5 + y^6$.

**(16.)** $(a + 2)^4 = a^4 + 8a^3 + 24a^2 + 32a + 16$.

**(17.)** $(a + 2b)^3 = a^3 + 6a^2b + 12ab^2 + 8b^3$.

**(18.)** $(3x + 5)^3 = 27x^3 + 135x^2 + 225x + 125$.

**(19.)** $(a^{\frac{2}{3}} + b^{\frac{1}{2}})^4 = (a^{\frac{2}{3}})^4 + 4(a^{\frac{2}{3}})^3b^{\frac{1}{2}} + 6(a^{\frac{2}{3}})^2(b^{\frac{1}{2}})^2 + 4a^{\frac{2}{3}}(b^{\frac{1}{2}})^3 + (b^{\frac{1}{2}})^4 = a^{\frac{8}{3}} + 3a^2b^{\frac{1}{2}} + 6a^{\frac{4}{3}}b + 4a^{\frac{2}{3}}b^{\frac{3}{2}} + b^2$.

**(20.)** $(m^{\frac{1}{2}} + b^{\frac{1}{3}})^6 = (m^{\frac{1}{2}})^6 + 6(m^{\frac{1}{2}})^5b^{\frac{1}{3}} + 15(m^{\frac{1}{2}})^4(b^{\frac{1}{3}})^2 + 20(m^{\frac{1}{2}})^3(b^{\frac{1}{3}})^3 + 15(m^{\frac{1}{2}})^2(b^{\frac{1}{3}})^4 + 6(m^{\frac{1}{2}})(b^{\frac{1}{3}})^5 + (b^{\frac{1}{3}})^6 = m^3 + 6m^{\frac{5}{2}}b^{\frac{1}{3}} + 15m^2b^{\frac{2}{3}} + 20m^{\frac{3}{2}}b + 15mb^{\frac{4}{3}} + 6m^{\frac{1}{2}}b^{\frac{5}{3}} + b^2$.

(Page 153, ***201.***)

**(21.)** $(a - b)^6 = a^6 - 6a^5b + 15a^4b^2 - 20a^3b^3 + 15a^2b^4 - 6ab^5 + b^6$.

**(22.)** $(1 - \frac{1}{2}a)^4 = 1 - 2a + \frac{3a^2}{2} - \frac{a^3}{2} + \frac{a^4}{16}$.

**(23.)** $(1 - 2x)^5 = (1)^5 - 5(1)^4(2x) + 10(1)^3(2x)^2 - 10(1)^2(2x)^3 + 5(1)(2x)^4 - (2x)^5 = 1 - 10x + 40x^2 - 80x^3 + 80x^4 - 32x^5$.

**(24.)** $(a^2 - a^{\frac{1}{2}})^4 = (a^2)^4 - 4(a^2)^3a^{\frac{1}{2}} + 6(a^2)^2(a^{\frac{1}{2}})^2 - 4(a^2)(a^{\frac{1}{2}})^3 + (a^{\frac{1}{2}})^4 = a^8 - 4a^{\frac{13}{2}} + 6a^5 - 4a^{\frac{7}{2}} + a^2$.

## EVOLUTION.

### (Page 154, *202.*)

**(1–5.)** $\sqrt{63504} = 252$. $\sqrt{15876} = 126$. $\sqrt{44100} = 210$. $\sqrt{200430649} = 14157$. $\sqrt{41616} = 204$.

**(6–15.)** $\sqrt{25a^{\frac{1}{3}}b^{\frac{2}{3}}c^{-4}} = \pm 5a^{\frac{1}{6}}b^{\frac{1}{3}}c^{-2}$. $\sqrt{16a^2b} = \pm 4ab^{\frac{1}{2}}$. $\sqrt{121x^{-4}y^{-\frac{4}{3}}} = \pm 11x^{-2}y^{-\frac{2}{3}}$. $\sqrt{169ay^{\frac{1}{2}}} = \pm 13a^{\frac{1}{2}}y^{\frac{1}{4}}$. $\sqrt{81m^5n^6} = \pm 9m^{\frac{5}{2}}n^3$. $\sqrt{\frac{36a^6x^4}{4x^2y^2}} = \pm 3\frac{a^3x}{y}$. $\sqrt[3]{125xy^2} = 5x^{\frac{1}{3}}y^{\frac{2}{3}}$. $\sqrt[3]{8x^{-2}y^{-3}} = \frac{2}{x^{\frac{2}{3}}y}$. $\sqrt[3]{-\frac{64x^3}{27a^6b^3}} = -\frac{4x}{3a^2b}$. $\sqrt[9]{-\frac{x^9y^{18}}{a^9b^9}} = -a^{-1}b^{-1}xy^2$.

### (Page 155, *204.*)

**(16–21.)** $\sqrt{36a^{-\frac{1}{2}}b^3} = \pm 6a^{-\frac{1}{4}}b^{\frac{3}{2}}$. $\sqrt[3]{-1728x^3y^{-1}} = -12xy^{-\frac{1}{3}}$. $\sqrt[4]{64a^{12}x^{-\frac{1}{3}}y^4} = \pm 4a^3x^{-\frac{1}{12}}y$. $\sqrt[5]{-32a^{20}c^{\frac{2}{5}}x} = -2a^4c^{\frac{2}{25}}x^{\frac{1}{5}}$. $\sqrt[3]{\frac{343a^2x^6}{27b^3y^3}} = \frac{7a^{\frac{2}{3}}x^2}{3by}$. $\sqrt[6]{\frac{64a^3x^2}{729b^{12}y^6}} = \frac{2a^{\frac{1}{2}}x^{\frac{1}{3}}}{3b^2y}$.

### (Page 156, *206.*)

**(22, 23.)** $\sqrt[3]{125 \times 8} = \sqrt[3]{125} \times \sqrt[3]{8}$. $\sqrt[3]{a^{-n}x^{-\frac{m}{n}}} = \sqrt[3]{a^{-n}} \times \sqrt[3]{x^{-\frac{m}{n}}}$.

**(24.)** $\sqrt{m^2x^2 - 2m^2xy + m^2y^2} = m(x - y)$.

**(25.)** $\sqrt{4a^2y^4 + 8a^2y^2 + 4a^2} = 2a(y^2 + 1)$.

### (Page 157, *207.*)

**(1.)** $\sqrt{9x^6 - 12x^5 + 10x^4 - 10x^3 + 5x^2 - 2x + 1} = 3x^3 - 2x^2 + x - 1$.

**(2.)** $\sqrt{4x^2y^2 + 12x^2y + 9x^2 - 30xy^2 - 20xy^3 + 25y^4} = 5y^2 - 2xy - 3x$.

**(3.)** $\sqrt{6x^3 + 7x^2 + 9x^4 + 1 + 2x} = 1 + x + 3x^2$.

**(4.)** $\sqrt{24x + 16 + 25x^2 + x^6 + 10x^4 + 4x^5 + 20x^3} = x^3 + 2x^2 + 3x + 4$.

**(5.)** $\sqrt{\frac{1051x^2}{25} - \frac{6x}{5} - \frac{14x^3}{5} + 9 + 49x^4} = 7x^2 - \frac{x}{5} + 3$.

(6.) $\sqrt{x^3+4x^{\frac{3}{2}}y^{\frac{5}{4}}+6x^{\frac{3}{2}}z^{\frac{7}{3}}+4y^{\frac{5}{2}}+12y^{\frac{5}{4}}z^{\frac{7}{3}}+9z^{\frac{14}{3}}}=x^{\frac{3}{2}}+2y^{\frac{5}{4}}+3z^{\frac{7}{3}}.$

(7.)

$$\sqrt{\frac{9a^{2m-2}c^2}{4d^{6p}}-\frac{3a^{m+n-1}b^{2n-1}c}{d^{3p-3}}-\frac{2^8a^{m-1}b^xc}{d^{3p}}+a^{2n}b^{4n-2}d^6+\frac{2^9}{3}a^nb^{x+2n-1}d^3+\frac{2^{16}b^{2x}}{9}}$$
$$=\frac{3a^{m-1}c}{2d^{3p}}-a^nb^{2n-1}d^3-\frac{2^8b^x}{3}.$$

## (Page 161, *209.*)

**(1–11.)** $\sqrt{9025}=95.$ $\sqrt{106929}=327.$ $\sqrt{387420489}=19683.$ $\sqrt{4782969}=2187.$ $\sqrt{943042681}=30709.$ $\sqrt{49042009}=7003.$ $\sqrt{1061326084}=32578.$ $\sqrt{.4}=.632455+.$ $\sqrt{.35}=.5916+.$ $\sqrt{7}=2.6457+.$ $\sqrt{500}=22.36+.$

**(12–15.)** $\sqrt{\frac{112}{175}}=.8.$ $\sqrt{\frac{81}{144}}=\frac{3}{4}.$ $\sqrt{\frac{3}{5}}=.7746,$ nearly. $\sqrt{\frac{5}{16}}=.559+.$

## (Page 165, *211.*)

**(1.)** $\sqrt[3]{-135x^2-125+27x^3+225x}=3x-5.$

**(2.)** $\sqrt[3]{108z^5x-27z^6-90z^4x^2+8x^6+48zx^5-80z^3x^3+60z^2x^4}=2x^2+4zx-3z^2.$

**(3.)** $\sqrt[3]{x^6+6x^5-40x^3+96x-64}=x^2+2x-4.$

**(4.)** $\sqrt[3]{-99x^3-9x^5+x^6+64-144x+156x^2+39x^4}=x^2-3x+4.$

**(5.)** $\sqrt[3]{27a^6-135a^5x+279a^4x^2-305a^3x^3+186a^2x^4-60ax^5+8x^6}=3a^2-5ax+2x^2.$

**(6.)** $\sqrt[3]{1-x^3}=1-\frac{1}{3}x^3-\frac{1}{9}x^6-\frac{5}{81}x^9-,$ etc.

**(7.)** $\sqrt[3]{x^{\frac{3}{2}}-3x^{\frac{4}{3}}+3x^{\frac{7}{6}}+3x^{\frac{5}{4}}+2x-3x^{\frac{5}{6}}-6x^{\frac{13}{12}}+3x^{\frac{11}{12}}+x^{\frac{3}{4}}}=x^{\frac{1}{2}}-x^{\frac{1}{3}}+x^{\frac{1}{4}}.$

## (Page 169, *212.*)

**(1–4.)** $\sqrt[3]{12812904}=234.$ $\sqrt[3]{1367631}=111.$ $\sqrt[3]{127263527}=503.$ $\sqrt[3]{158252632929}=5409.$

**(5—9.)** $\sqrt[3]{8.5} = 2.0408+$. $\sqrt[3]{.4} = .7368+$. $\sqrt[3]{.25} = .62996+$. $\sqrt[3]{28.25} = 3.045+$. $\sqrt[3]{102.875} = 4.685+$.

**(10—15.)** $\sqrt[3]{12} = 2.289+$. $\sqrt[3]{9} = 2.08+$. $\sqrt[3]{\frac{5}{9}} = .822+$. $\sqrt[3]{15\frac{2}{3}} = 2.5022+$. $\sqrt[3]{3\frac{4}{5}} = 1.56+$. $\sqrt[3]{465\frac{31}{64}} = 7\frac{3}{4}$.

## (Page 174, *213.*)

**(1.)** $(81x^8 - 216x^7 + 336x^5 - 56x^4 - 224x^3 + 64x + 16)^{\frac{1}{4}} = 3x^2 - 2x - 2$.

**(2.)** $(729 - 2916x^2 + 4860x^4 - 4320x^6 + 2160x^8 - 576x^{10} + 64x^{12})^{\frac{1}{6}} = 3 - 2x^2$.

**(3, 4.)** $\sqrt[6]{11390625} = 15$. $\sqrt[8]{214358881} = 11$.

## (Pages 174, 175, *214, 215.*)

**(1.)** $\sqrt[4]{a^4 + 4a^3bx + 6a^2b^2x^2 + 4ab^3x^3 + b^4x^4} = a + bx$.

**(2.)** $(x^{-20} + 15x^{-16} - 5x^{-14} + 90x^{-12} - 60x^{-10} + 282x^{-8} - 252x^{-6} + 505x^{-4} - 496x^{-2} + 495 - 495x^2 + 275x^4 - 80x^6 + 15x^8 - x^{10})^{\frac{1}{5}} = x^{-4} + 3 - x^2$.

**(3.)** $\sqrt[5]{32x^5 - 80x^4 + 80x^3 - 40x^2 + 10x - 1} = 2x - 1$.

**(4—7.)** $\sqrt[5]{39} = 2.08+$. $\sqrt[7]{108} = 1.952+$. $\sqrt[5]{33554432} = 32$. $\sqrt[11]{3} = 1.031$.

---

# REDUCTION OF RADICALS.

## (Pages 178, 180, *217, 218.*)

**(1—10.)** $\sqrt[3]{54a^4b^3c^2} = 3ab\sqrt[3]{2ac^2}$. $\sqrt{192a^3x^2y} = 8ax\sqrt{3ay}$. $\sqrt{45a^4b^6} = 3a^2b^3\sqrt{5}$. $\sqrt[3]{250a^7x^3y^2} = 5a^2x\sqrt[3]{2ay^2}$. $\sqrt{243a^3b^2c} = 9ab\sqrt{3ac}$. $\sqrt[6]{192a^7bc^{12}} = 2ac^2\sqrt[6]{3ab}$. $\sqrt[5]{x^{10}y^{-5}z^{5m+1}} = x^2y^{-1}z^m\sqrt[5]{z}$. $(56a^{6m-1}x^{3-n})^{\frac{1}{3}} = 2a^{3m}x\left[\frac{7}{ax^n}\right]^{\frac{1}{3}}$. $(1215x^{\frac{10}{7}}y^8)^{\frac{1}{5}} = 3x^{\frac{2}{7}}y(5y^3)^{\frac{1}{5}}$.

**(11—17.)** $\sqrt[3]{x^4 - a^2x^3} = x\sqrt[3]{x - a^2}$. $\sqrt[m]{a^{2m+1}b^{1-m}} = \frac{a^2}{b}\sqrt[m]{ab}$. $\sqrt[3]{(a+b)^2(a^2-b^2)} = (a+b)\sqrt[3]{a-b}$. $5\sqrt{147a^3x^2} = 35ax\sqrt{3a}$. $(a+b)^{-1}\sqrt{a^2 + 2ab + b^2} = 1$. $2a^{-2}\sqrt{12a^{-5}b^3} = 4a^{-4}b\sqrt{\frac{3b}{a}}$. $11\sqrt{44x^2 - 40x^3y} = 22x\sqrt{11 - 10xy}$.

**(18—24.)** $\sqrt{\frac{27}{5}} = \frac{3}{5}\sqrt{15}$. $\sqrt{\frac{50}{7}} = \frac{5}{7}\sqrt{14}$. $\sqrt{\frac{1}{7}} = \frac{1}{7}\sqrt{7}$. $\sqrt{\frac{56}{25}} = \frac{2}{5}\sqrt{14}$. $\sqrt[3]{\frac{1}{11}} = \frac{1}{11}\sqrt[3]{121}$. $\frac{a^2 - x^2}{27}\sqrt{\frac{18}{a^2 - 2ax + x^2}} = \frac{a+x}{9}\sqrt{2}$. $\frac{1}{a^2 + x^2}\sqrt{(a^4 - x^4)(a^2 + x^2)} = \sqrt{a^2 - x^2}$.

## (Page 181, *219.*)

**(1—6.)** $\sqrt[6]{4a^2} = \sqrt[3]{2a}$. $\sqrt[4]{36a^2b^2} = \sqrt{6ab}$. $\sqrt[10]{32a^5b^5} = \sqrt{2ab}$. $\sqrt[12]{16a^4x^2y^{2m}z^{4n-4}} = \sqrt[6]{4a^2xy^mz^{2n-2}}$. $\sqrt[8]{256a^{12}} = \sqrt{4a^3} = 2a\sqrt{a}$. $\sqrt[6]{a^3 - 3a^2b + 3ab^2 - b^3} = \sqrt{a - b}$.

## (Page 182, *220.*)

**(1—6.)** $3xy^2 = \sqrt[3]{27x^3y^6}$. $1 - x^2 = \sqrt{1 - 2x^2 + x^4}$. $3a^{\frac{1}{2}} - 2 = \sqrt{9a - 12a^{\frac{1}{2}} + 4}$. $\frac{1}{3} = \sqrt{\frac{1}{9}} = \sqrt[3]{\frac{1}{27}}$. $\frac{x^3}{3} = \sqrt[3]{\frac{x^9}{27}}$. $2a - 1 = \sqrt[3]{8a^3 - 12a^2 + 6a - 1}$

## (Page 183, *221.*)

**(2.)** $2x\sqrt[3]{3x^{-2}} = \sqrt[3]{24x}$. $\frac{1}{2}\sqrt{12} = \sqrt{3}$. $\frac{1}{3}\sqrt[3]{54} = \sqrt[3]{2}$. $\frac{1}{5}\sqrt[3]{200} = \sqrt[3]{\frac{8}{5}}$. $2a\sqrt[3]{3ax} = \sqrt[3]{24a^4x}$. $(a^2 - b^2)\sqrt{a - b} = \sqrt{(a-b)^3(a+b)^2}$ $\frac{3}{2}ax^{-\frac{1}{2}}\sqrt[3]{\frac{2}{3}ax} = \sqrt[3]{\frac{9}{4}a^4x^{-\frac{1}{2}}}$. $y(2r - y)^{\frac{1}{2}} = \sqrt{2ry - y^2}$. $x^3(2 - x)^{\frac{3}{2}} = (2x^2 - x^3)^{\frac{3}{2}}$. $\frac{1}{x^2}(ax - x^2)^{\frac{2}{3}} = \left(\frac{a}{x^2} - \frac{1}{x}\right)^{\frac{2}{3}}$.

## (Page 185, *222.*)

**(1.)** $\sqrt{2}$, $\sqrt[3]{3}$, $\sqrt[4]{5} = \sqrt[24]{4096}$, $\sqrt[24]{6561}$, $\sqrt[24]{15625}$

**(2.)** $\sqrt[3]{2}$, $\sqrt{3} = \sqrt[6]{4}$, $\sqrt[6]{27}$.

**(3.)** $a^2$, $b^{\frac{1}{2}} = \sqrt{a^4}$, $\sqrt{b}$.

**(4.)** $\sqrt[4]{a}$, $\sqrt[6]{5b}$, $\sqrt[8]{6c} = \sqrt[24]{a^6}$, $\sqrt[24]{625b^4}$, $\sqrt[24]{216c^3}$.

**(5.)** $\sqrt{2a^2x}$, $\sqrt[3]{2ax^2} = \sqrt[6]{8a^6x^3}$, $\sqrt[6]{4a^2x^4}$.

**(6.)** $(ax)^{\frac{1}{3}}$, $(b^2x)^{\frac{2}{5}}$, $(3a^2)^{\frac{1}{2}} = \sqrt[30]{a^{10}x^{10}}$, $\sqrt[30]{b^{24}x^{12}}$, $\sqrt[30]{3^{15}a^{30}}$.

(**7.**) $\sqrt[2n]{a}, \sqrt[2m]{a} = \sqrt[2mn]{a^m}, \sqrt[2mn]{a^n}$.

(**8.**) $2\sqrt{a},\ 3\sqrt[3]{b} = 2\sqrt[6]{a^3},\ 3\sqrt[6]{b^2}$.

(**9.**) $4\sqrt{2a^2x},\ \sqrt[6]{3xy},\ 10\sqrt[3]{x^2y} = 4\sqrt[6]{8a^6x^3},\ \sqrt[6]{3xy},\ 10\sqrt[6]{x^4y^2}$.

(**10.**) $(2-x)^{\frac{1}{2}},\ (2-x)^{\frac{1}{3}} = \sqrt[6]{(2-x)^3},\ \sqrt[6]{(2-x)^2}$.

(PAGE 187, ***223.***)

$\frac{2a\sqrt{3x}}{\sqrt{5x}} = \frac{2a}{5}\sqrt{15}$. $\frac{1}{3\sqrt{b}} = \frac{\sqrt{b}}{3b}$. $\frac{\sqrt{x^3}}{\sqrt{xy}} = \frac{x\sqrt{y}}{y}$. $\frac{2x}{\sqrt[3]{ax}} = \frac{2x\sqrt[3]{a^2x^2}}{ax}$. $\frac{1}{\sqrt{2}} = \frac{\sqrt{2}}{2} = \frac{1}{2}\sqrt{2}$. $\frac{2}{\sqrt{3}} = \frac{2\sqrt{3}}{3} = \frac{2}{3}\sqrt{3}$. $\frac{\sqrt{5}}{\sqrt{14}} = \frac{\sqrt{70}}{14}$. $\frac{a-x}{\sqrt{a+x}} = \frac{\sqrt{(a^2-x^2)(a-x)}}{a+x}$.

(PAGE 189, ***224.***)

(**1—4.**) $\frac{\sqrt{a}}{2-\sqrt{a}} = \frac{2\sqrt{a}+a}{4-a}$. $\frac{3}{\sqrt{5}-\sqrt{2}} = \sqrt{5}+\sqrt{2}$. $\frac{4\sqrt{3}}{2\sqrt{3}+3\sqrt{2}} = 2\sqrt{6}-4$. $\frac{1-\sqrt{5}}{3+\sqrt{5}} = 2-\sqrt{5}$.

(**5, 6.**) $\frac{\sqrt{3}+1}{2-\sqrt{3}} = 5+3\sqrt{3}$. $\frac{\sqrt{20}+\sqrt{12}}{\sqrt{5}-\sqrt{3}} = 8+2\sqrt{15}$. $\frac{1}{x+\sqrt{x^2-1}} + \frac{1}{x-\sqrt{x^2-1}} = 2x$. $\frac{a+bx-\sqrt{a^2+b^2x^2}}{a-bx+\sqrt{a^2+b^2x^2}} = \frac{\sqrt{a^2+b^2x^2}-a}{bx}$.

(PAGE 190, ***225.***)

(**1.**) What factor will rationalize $1-\sqrt[3]{2a}$? $a^{\frac{3}{4}}+b^{\frac{3}{4}}$? $3\sqrt[3]{3}-\frac{1}{3}\sqrt[3]{2}$? $a^{\frac{1}{2}}-b^{\frac{2}{3}}$? What are the rationalized products?

*Answers.* $1+\sqrt[3]{2a}+\sqrt[3]{4a^2}$, $\sqrt[4]{a^9}-\sqrt[4]{a^6b^3}+\sqrt[4]{a^3b^6}-\sqrt[4]{b^9}$, $\sqrt[3]{9}+\frac{1}{2}\sqrt[3]{6}+\frac{1}{3}\sqrt[3]{4}$, $a^{\frac{5}{2}}+a^2b^{\frac{2}{3}}+a^{\frac{3}{2}}b^{\frac{4}{3}}+ab^2+a^{\frac{1}{2}}b^{\frac{8}{3}}+b^{\frac{10}{3}}$, are the factors.

(PAGE 192, ***226.***)

(**1.**) What factors rationalize $\sqrt{5}+\sqrt{3}-\sqrt{2}$, and $\sqrt{3}+\sqrt{2}+1$? What are the rational products?

*Answers* The factors are $6\sqrt{2}-4\sqrt{3}-2\sqrt{30}$, and $2\sqrt{6}-2\sqrt{2}-4$. The products are $-24$, and $-8$

(**2.**) Show that $\frac{\sqrt[3]{x^2}}{\sqrt[3]{x^2}-\sqrt[3]{xy}+\sqrt[3]{y^2}} = \frac{x+\sqrt[3]{x^2y}}{x+y}$.

## ADDITION AND SUBTRACTION OF RADICALS.

(PAGE 193, ***227.***)

**(1–18.)** $3\sqrt{13}+8\sqrt{13}=11\sqrt{13}$. $3\sqrt{180}+4\sqrt{405}=54\sqrt{5}$. $8\sqrt[3]{54}+5\sqrt[3]{128}=44\sqrt[3]{2}$. $\sqrt{\frac{3}{4}}-\sqrt{\frac{1}{3}}=\frac{1}{6}\sqrt{3}$. $\sqrt{\frac{3}{4}}+\sqrt{\frac{1}{3}}+\frac{1}{2}\sqrt{3}=\frac{4}{3}\sqrt{3}$. $(2\sqrt{\frac{5}{3}}+\sqrt{60})-(\sqrt{15}-\sqrt{\frac{3}{5}})=\frac{28}{15}\sqrt{15}$. $\sqrt{45x^3}-\sqrt{80x^3}+\sqrt{5a^2x}=(a-x)\sqrt{5x}$. $\sqrt[3]{16a^3b}+\sqrt{4a^2b}-\sqrt{a^2b}-\sqrt[3]{54a^3b}=a\sqrt{b}-a\sqrt[3]{2b}$. $\sqrt[3]{\frac{27a^5x}{2b}}-\sqrt[3]{\frac{a^2x}{2b}}=(3a-1)\sqrt[3]{\frac{a^2x}{2b}}$. $4a\sqrt[3]{a^3b^4}+b\sqrt[3]{8a^6b}-\sqrt[3]{125a^6b^4}=a^2b\sqrt[3]{b}$. $\sqrt{\frac{ab^3}{c^2}}+\frac{1}{2c}\sqrt{(a^3b-4a^2b^2+4ab^3)}=\frac{a}{2c}\sqrt{ab}$. $m^{\frac{3}{4}}\sqrt{1+\left(\frac{n}{m}\right)^{\frac{1}{2}}}-n^{\frac{3}{4}}\sqrt{1+\left(\frac{m}{n}\right)^{\frac{1}{2}}}=(m^{\frac{1}{2}}-n^{\frac{1}{2}})\sqrt{m^{\frac{1}{2}}+n^{\frac{1}{2}}}$. $\sqrt{\frac{a+x}{a-x}}+\sqrt{\frac{a-x}{a+x}}=\frac{2a}{a^2-x^2}\sqrt{a^2-x^2}$. $\sqrt{27a^2c+54abc+27b^2c}+3\sqrt{3cx^2}=3(x+a+b)\sqrt{3c}$. $\sqrt{4a^5b^2-20a^3b^3+25ab^4}-7\sqrt{ac^6}=(2a^2b-5b^2-7c^3)\sqrt{a}$. $6\sqrt[6]{4a^2}+2\sqrt[3]{2a}+\sqrt[9]{8a^3}=9\sqrt[3]{2a}$. $\sqrt[4]{16}+\sqrt[3]{81}-\sqrt[3]{-512}+\sqrt[3]{192}-7\sqrt[6]{9}=10$. $\sqrt{\frac{a^2x-2ax^2+x^3}{a^2+2ax+x^2}}+\sqrt{\frac{a^2x+2ax^2+x^3}{a^2-2ax+x^2}}=2\left(\frac{a^2+x^2}{a^2-x^2}\right)\sqrt{x}$.

---

## MULTIPLICATION OF RADICALS.

(PAGE 195, ***230.***)

**(1–19.)** $\sqrt{8}\times\sqrt{2}=4$. $3\sqrt{2}\times 2\sqrt{3}=6\sqrt{6}$. $\sqrt{\frac{1}{3}}\times\sqrt{\frac{3}{4}}=\frac{1}{2}$. $4\sqrt[3]{12}\times 3\sqrt[3]{4}=24\sqrt[3]{6}$. $m\sqrt[a]{x}\times n\sqrt[b]{y}=mn\sqrt[ab]{x^by^a}$. $4\sqrt{3}\times 2\sqrt[3]{2}=8\sqrt[6]{108}$. $5a^{\frac{2}{3}}b^{\frac{1}{5}}\times 2a^{\frac{1}{3}}b^{\frac{4}{5}}=10ab$. $(a+b)^{\frac{1}{3}}\times(a+b)^{\frac{3}{4}}=(a+b)^{\frac{13}{12}}$. $\sqrt{7}\times\sqrt[3]{7}=\sqrt[6]{16807}$. $2\sqrt[3]{14}\times 3\sqrt[3]{4}=12\sqrt[3]{7}$. $\sqrt[6]{a^2b^2d^3}\times\sqrt{d}=d\sqrt[3]{ab}$. $\left(1+\frac{x^2}{y^2}\right)^{\frac{3}{2}}\times y^3=(y^2+x^2)^{\frac{3}{2}}$. $\sqrt{5abc^3}\times\sqrt[3]{2a^2bc^2}=ac^2\sqrt[6]{500ab^5c}$. $3a\sqrt[6]{b}\times 5b\sqrt[8]{2c}=15ab\sqrt[24]{8b^4c^3}$. $\sqrt[3]{2}\times\sqrt[6]{\frac{1}{3}}\times\sqrt[8]{3}=$

$\sqrt[24]{\frac{256}{3}}$. $\frac{a+b}{a-b} \times \sqrt{\frac{a-b}{a+b}} = \sqrt{\frac{a+b}{a-b}}$. $\frac{a-b}{a+b} \times \frac{\sqrt{ac}}{\sqrt{a^2-2ab+b^2}} = \frac{\sqrt{ac}}{a+b}$. $\sqrt[12]{\frac{a}{bc}} \times \sqrt[8]{\frac{a^m}{b}} = \sqrt[24]{\frac{a^{3m+2}}{b^5c^2}}$. $x^{\frac{m}{n}} \times x = x^{\frac{m+n}{n}}$

**(22—28.)** $(9+2\sqrt{10}) \times (9-2\sqrt{10}) = 41$. $(a^2 + a\sqrt[3]{b} + \sqrt[3]{b^2}) \times (a - \sqrt[3]{b}) = a^3 - b$. $(\sqrt[3]{x^2} - \sqrt[3]{xy} + \sqrt[3]{y^2})(\sqrt[3]{x} + \sqrt[3]{y}) = x + y$. $(5 - 8\sqrt{7}) \times (9 + 10\sqrt{3}) = 45 - 72\sqrt{7} + 50\sqrt{3} - 80\sqrt{21}$. $(x - \sqrt{xy} + y) \times (\sqrt{x} + \sqrt{y}) = x\sqrt{x} + y\sqrt{y} = x^{\frac{3}{2}} + y^{\frac{3}{2}}$. $(\sqrt{2} + 3\sqrt{\tfrac{1}{2}}) \times \tfrac{1}{10} = \tfrac{1}{2}\sqrt{\tfrac{1}{2}}$. $\sqrt[5]{2} \times 2 = \sqrt[5]{64}$. $(\sqrt{2} + \sqrt{3}) \times (2\sqrt{2} - \sqrt{3}) = 1 + \sqrt{6}$. $(\sqrt{a} - \sqrt{a-x}) \times (\sqrt{a} + \sqrt{a-x}) = x$.

---

## DIVISION OF RADICALS.

### (Page 199, *232.*)

**Many examples can be produced by assigning the quotient and one factor from the above. A few additional exercises are given.**

$\sqrt{54} \div \sqrt{6} = 3$. $\sqrt{160} \div 8 = 2\sqrt{5}$. $6\sqrt{28} \div 2\sqrt{7} = 6$. $x \div \sqrt{x} = \sqrt{x}$. $a \div \sqrt[3]{a} = a^{\frac{2}{3}}$. $\sqrt{\tfrac{1}{2}} \div \sqrt{\tfrac{1}{3}} = \tfrac{1}{2}\sqrt{6}$. $\tfrac{3}{5}\sqrt{\tfrac{1}{3}} \div \tfrac{1}{2}\sqrt{\tfrac{3}{5}} = \tfrac{2}{5}\sqrt{5}$. $4\sqrt[3]{12} \div 2\sqrt{3} = 2\sqrt[6]{\tfrac{16}{3}}$. $\sqrt{12} \div \sqrt[3]{24} = \sqrt[6]{3}$. $(a+b)^{\frac{1}{2}} \div (a^2-b^2)^{\frac{1}{3}} = \sqrt[6]{\frac{a+b}{(a-b)^2}}$. $\sqrt[10]{144a^3x^2} \div \sqrt[5]{6ax} = \sqrt[10]{4a}$. $\sqrt{ab^2 - b^2c} \div \sqrt{a-c} = b$. $(\sqrt{72} + \sqrt{32} - 4) \div \sqrt{8} = 5 - \sqrt{2}$. $(x^4+1) \div (x^2 + x\sqrt{2} + 1) = x^2 - x\sqrt{2} + 1$. $(a - b^2) \div \sqrt[4]{a^3} + \sqrt{ab} + \sqrt[4]{ab} + \sqrt{b^3} = \sqrt[4]{a} - \sqrt{b}$. $\sqrt[6]{\frac{y}{x}} \div \sqrt{\frac{ay}{x}} = \sqrt[3]{\frac{x}{\sqrt{ay}}}$. $x\sqrt{\frac{a-b}{a+b}} \div y\sqrt{\frac{a-b}{a+b}} = \frac{x}{y}$. $(a+b)\sqrt{a^2-1} \div (a-b)\sqrt{(a+1)^2} = \frac{a+b}{a-b}\sqrt{\frac{a-1}{a+1}}$.

## INVOLUTION OF RADICALS.

(PAGE 201, *233.*)

**(1—7.)** $(\frac{1}{2}\sqrt{2ax})^3 = \frac{1}{4}ax\sqrt{2ax}$. $(3\sqrt[3]{3})^2 = 9\sqrt[3]{9}$. $(-\sqrt[3]{a^2})^4 = a^2\sqrt[3]{a^2}$. $(a-\sqrt{b})^3 = a^3 - 3a^2\sqrt{b} + 3ab - b\sqrt{b}$, $(3+\sqrt{5})^2 = 14 + 6\sqrt{5}$. $\left\{\left(\frac{1+m}{2}\right)^{\frac{1}{2}} + \left(\frac{1-m}{2}\right)^{\frac{1}{2}}\right\}^2 = 1 + (1-m^2)^{\frac{1}{2}}$. $(3\sqrt[3]{5})^2 = 9\sqrt[3]{25}$.

---

## EVOLUTION OF RADICALS.

(PAGE 202, *235.*)

**(1—12.)** $\sqrt{9\sqrt[3]{3}} = 3\sqrt[6]{3}$. $\sqrt{(5a^2 - 3x^2)^{\frac{3}{2}}} = \sqrt{5a^2 - 3x^2}$. $\sqrt[3]{\frac{1}{8}a^3b} = \frac{1}{2}a\sqrt[3]{b}$. $\sqrt[n]{\sqrt[m]{a^nx^2}} = a^{\frac{1}{m}}x^{\frac{2}{mn}}$. $\sqrt[4]{\sqrt[3]{16x^5}} = \sqrt[3]{2x\sqrt[4]{x}}$. $\sqrt[4]{625\sqrt[3]{5}} = 5\sqrt[12]{5}$. $\sqrt[3]{\frac{a}{5}\sqrt{\frac{a}{5}}} = \frac{1}{5}\sqrt{5a}$. $\sqrt[4]{2\sqrt{a^2-x^2}} = \sqrt[8]{4(a^2-x^2)}$. $\sqrt[m]{5(a+b)^{\frac{1}{2}}} = 5^{\frac{1}{m}}(a+b)^{\frac{1}{2m}}$. $(3\sqrt[3]{5})^{\frac{1}{2}} = \sqrt[6]{135}$. $\sqrt[3]{(m+n)\sqrt{m+n}} = \sqrt{m+n}$. $\sqrt[5]{\sqrt{32x^5}} = \sqrt{2x}$.

(PAGE 205, *237.*)

**(1—7.)** $\sqrt{7+4\sqrt{3}} = 2 + \sqrt{3}$. $\sqrt{19+8\sqrt{3}} = 4 + \sqrt{3}$. $\sqrt{a^2+b+2a\sqrt{b}} = a + \sqrt{b}$. $\sqrt{\sqrt{32}-\sqrt{24}} = \sqrt[4]{18} - \sqrt[4]{2}$. $\sqrt{\sqrt{27}-2\sqrt{6}} = \sqrt[4]{12} - \sqrt[4]{3}$. $\sqrt{2x+2\sqrt{x^2-1}} = \sqrt{x+1} + \sqrt{x-1}$. $\sqrt{3\sqrt{6}-4\sqrt{3}} = \sqrt[4]{24} - \sqrt[4]{6}$.

## IMAGINARIES.

### (Page 207, *241.*)

**(1—6.)** $(2\sqrt{-3}) \times 3\sqrt{-5} = -6\sqrt{15}$. $(3\sqrt{-2}) \times (\sqrt{-8}) = -12$. $\sqrt{-a^2} \times \sqrt{-b^4} \times \sqrt{-c^2} = -ab^2c\sqrt{-1}$, or $-\sqrt{-a^2b^4c^2}$. $(2+\sqrt{-5}) \times (2+\sqrt{-5}) = -1+4\sqrt{-5}$. $(3+\sqrt{-2}) \times (3-\sqrt{-2}) = 11$. $\{x+\frac{1}{2}(1-\sqrt{-3})\} \times \{(x-1)(x+\frac{1}{2}+\frac{1}{2}\sqrt{-3})\} = x^3-1$.

**(7—10.)** $(6\sqrt{-40}) \div (2\sqrt{-10}) = 6$. $(a\sqrt{-1}) \div (b\sqrt{-1}) = \frac{a}{b}$. $(5-\sqrt{-2}) \div (1+\sqrt{-2}) = 1-2\sqrt{2}\sqrt{-1}$. $1 \div \sqrt{-1} = -\sqrt{-1}$.

**(11.)** Show that $\dfrac{a+\sqrt{-b}}{a-\sqrt{-b}} + \dfrac{a-\sqrt{-b}}{a+\sqrt{-b}} = \dfrac{2(a^2-b)}{a^2+b}$.

*PART II.*

# ALGEBRA.

---

## CLEARING EQUATIONS OF FRACTIONS.

(PAGE 213, ***14.***)

**(1.)** $\frac{y}{3} - \frac{y}{5} + \frac{y}{10} = \frac{3y - 2}{5}$. $\therefore$ $10y - 6y + 3y = 18y - 12$.

**(2).** $\frac{z}{2} - \frac{3z}{4} + \frac{5z}{2} = \frac{5}{6}$. $\therefore$ $6z - 9z + 30z = 10$.

**(3.)** $\frac{x-1}{4} - \frac{x}{3} = \frac{2x+3}{8} - 5x$. $\therefore$ $6x - 6 - 8x = 6x + 9 - 120x$.

**(4.)** $\frac{y-2}{3} - 2y = \frac{3y}{5} - \frac{3y-5}{10}$. $\therefore$ $10y - 20 - 60y = 18y - 9y + 15$.

**(5.)** $\frac{z}{ac} - \frac{z}{b} + \frac{z}{c^2} = 2z - \frac{3z - b}{bc}$. $\therefore$ $bcz - ac^2z + abz = 2abc^2z - 3acz + abc$.

**(6.)** $\frac{y}{7} - \frac{y+2}{21} = 4y + \frac{y-1}{14}$. $\therefore$ $6y - 2y - 4 = 168y + 3y - 3$.

**(7.)** $\frac{3z}{2m} - \frac{2z+1}{3am^2} + \frac{z-2}{am} = 2 - z$. $\therefore$ $9amz - 4z - 2 + 6mz - 12m = 12am^2 - 6am^2z$.

**(8.)** $\frac{y}{m-n} + y - \frac{2}{(m+n)^2} = \frac{m+n}{m-n}$. $\therefore$ $m^2y + 2mny + n^2y + m^3y + m^2ny - mn^2y - n^3y - 2m + 2n = (m+n)^3$.

**(9.)** $\frac{2-a}{x+a} = \frac{3a-1}{x-a}$. $\therefore$ $2x - 2a - ax + a^2 = 3ax + 3a^2 - x - a$.

**(10.)** $3 - \frac{2y}{a-b} = \frac{y}{2a+b}$. $\therefore$ $6a^2 - 3ab - 3b^2 - 4ay - 2by = ay - by$.

## TRANSPOSITION.

(PAGE 216, ***16.***)

In the following examples transpose the terms containing the unknown quantity to the first member, and the known terms to the second.

(**1.**) $3x - 4 + 2x + 5 = 5x - 1 - 6x + 8.$

$$\therefore\ 3x + 2x - 5x + 6x = 4 - 5 - 1 + 8.$$

(**2.**) $5y - 10 - 2y + 6 - y = 4 - 3y - 1 + 2y.$

$$\therefore\ 5y - 2y - y + 3y - 2y = 4 - 1 - 6 + 10.$$

(**3.**) $3ac + 2ax - 5mx + a^2c = 3a^2c^2 - 2mx - 3ax - ac.$

$$\therefore\ 2ax - 5mx + 2mx + 3ax = 3a^2c^2 - ac - a^2c - 3ac.$$

(**4.**) $\frac{3y}{5} - \frac{5}{6} + \frac{2y}{3} + \frac{2}{3} = \frac{5y}{3} - \frac{2y}{7} + \frac{11}{5}.$

$$\therefore\ \frac{3y}{5} + \frac{2y}{3} - \frac{5y}{3} + \frac{2y}{7} = \frac{11}{5} - \frac{2}{3} + \frac{5}{6}.$$

(**5.**) $\frac{2ax}{4} - 2am + \frac{3x}{5a} + 4ac^2 = \frac{x}{5} - 2ac^2 - \frac{2x}{a}.$

$$\therefore\ \frac{2ax}{4} + \frac{3x}{5a} - \frac{x}{5} + \frac{2x}{a} = 2am - 4ac^2 - 2ac^2.$$

---

## SOLUTION OF SIMPLE EQUATIONS WITH ONE UNKNOWN QUANTITY.

(PAGE 220, ***20, 21.***)

(**1.**) $\frac{x-1}{4} + \frac{11x-2}{3} = \frac{1-x}{4} + \frac{59}{3}.\ \ \therefore\ x = 5.$

(**2.**) $19x + 13 = 59 - 4x.\ \ \therefore\ x = 2.$

(**3.**) $3x + 4 - \frac{x}{3} = 46 - 2x.\ \ \therefore\ x = 9.$

(**4.**) $\frac{3x+1}{2} - \frac{2x}{3} = 10 + \frac{x-1}{6}.\ \ \therefore\ x = 14.$

(**5.**) $45y - 42 - 3y = 96 - 8y.\ \ \therefore\ = \frac{138}{50} = 2\frac{19}{25}.$

(6.) $\frac{2y+1}{2} = \frac{7y+5}{8}$. $\therefore y = 1$.

(7.) $\frac{x-3}{4} + \frac{x-4}{3} = \frac{x-5}{2} + \frac{x+1}{8}$. $\therefore x = 7$.

(8.) $\frac{41}{60} + \frac{3x-8}{5} + \frac{2x-5}{6} = \frac{x-3}{4} + \frac{5x+6}{15}$. $\therefore x = 4$.

(9.) $5(3+y) = 7(2y-3)$. $\therefore y = 4$.

(10.) $6(x-1) = 3\frac{1}{3} - \frac{2x}{3} + 4x$. $\therefore x = 3\frac{1}{2}$.

(11.) $\frac{2x-9}{3} = \frac{x}{4} - \frac{5x+8}{6}$. $\therefore x = 1\frac{1}{3}$.

(12.) $\frac{2y-11}{3} + \frac{19-2y}{2} = 2y$. $\therefore y = 2\frac{1}{2}$.

(13.) $\frac{3z-14}{3} = \frac{3z-7}{5} + \frac{25-4z}{9}$. $\therefore z = 4$.

(14.) $\frac{10y-427}{19} = \frac{40-y}{8} + \frac{2y+5}{13}$. $\therefore y = 56$.

(15.) $\frac{5x-7}{2} = \frac{2x+7}{3} + 3x - 14$. $\therefore x = 7$.

(16.) $\frac{2x-9}{27} + \frac{x}{18} = 8\frac{1}{3} - x + \frac{x-3}{4}$. $\therefore x = 9$.

(17.) $\frac{3x-11}{4} + 14\frac{3}{4} = 4x + \frac{28-9x}{8}$. $\therefore x = 4$.

(18.) $x + \frac{11-x}{3} = \frac{26-x}{2}$. $\therefore x = 8$.

(19.) $\frac{2x-1}{3} - \frac{3x-2}{4} = \frac{5x-4}{6} - \frac{7x+6}{12}$. $\therefore x = 4$.

(20.) $\frac{x+1}{2} - \frac{5-x}{4} = 14 - \frac{x+2}{3}$. $\therefore x = 13$.

(21.) $4\frac{1}{2} - \frac{9x}{2} - 2x = \frac{3x}{4} - 7x - \frac{2x}{5} + \frac{1}{3}x - 5$. $\therefore x = 51\frac{9}{11}$.

(22.) $\frac{3x}{52} = 2\left(1 - \frac{30-x}{19}\right)$. $\therefore x = 24\frac{16}{47}$.

(23.) $\frac{1}{2}(x-1) + \frac{1}{3}(x-2) = \frac{1}{4}(x+3) + \frac{1}{6}(x+4) + 1$. $\therefore x = 8\frac{3}{5}$.

(**24.**) $\frac{1}{7}x - \frac{1}{11}(x-5)+7 = x - \left(\frac{2x}{77} - 1\right)$. $\therefore x = 7$.

(**25.**) $\frac{1}{4}(7x + 9) - \left(x - \frac{2x-1}{9}\right) = 7$. $\therefore x = 5$.

(**26.**) $\frac{1}{4}(7 + 9x) - \left(1 - \frac{2-x}{9}\right) = 7x$. $\therefore x = \frac{1}{5}$.

(**27.**) $\frac{dx}{c} + 3ab = \frac{x}{a} + 1$. $\therefore x = \frac{ac(1-3ab)}{ad-c}$.

(**28.**) $\frac{5}{6}ab + \frac{4}{5}ac - \frac{2}{3}cx = \frac{3}{4}ac + 2ab - 6cx$. $\therefore x = \frac{(70b-3c)a}{320c}$.

(**29.**) $\frac{f^2x}{cg} - \frac{a^2}{f} + cx = \frac{hx}{g} - c + (a+c)x$. $\therefore x = \frac{(a^2-cf)cg}{(f^2-ch-acg)f}$.

(**30.**) $\frac{ace}{d} + \frac{(a+b)^2x}{a} + bx = ae + 3bx$. $\therefore x = \frac{a^2e(d-c)}{(a^2+b^2)d}$.

(**31.**) $\frac{bx}{2b-a} - \frac{(3bc+ad)x}{2ab(a+b)} - \frac{5ab}{3c-d} = \frac{(3bc-ad)x}{2ab(a-b)} - \frac{5a(2b-a)}{a^2-b^2}$.
$\therefore x = \frac{5a(2b-a)}{3c-d}$.

(**32.**) $\frac{a-3x}{4a} - \frac{7a+5x}{6b} + 3 + \frac{9x}{4} = -\frac{x}{ab} - \frac{5x}{6b}$. $\therefore x = \frac{39ab - 14a^2}{-27ab + 9b - 12}$.

(**33.**) $-\frac{3abc}{a-b} + \frac{a^2b^2}{(a-b)^3} - \frac{(2a-b)b^2x}{a(a-b)^2} = \frac{bx}{a} - 3cx$. $\therefore x = \frac{ab}{a-b}$.

(**34.**) $cd - \frac{a^2x}{b-c} + bx + ac = 0$. $\therefore x = \frac{c(a+d)(b-c)}{a^2-b(b-c)}$.

(**35.**) $2a^2b^2c - ab^2x - 2ab^3c - abc^2d + 3a^3x = (3a^2b - b^3)x - b^2c^2d$.
$\therefore x = \frac{bc^2d - 2ab^2c}{3a^2-b^2}$.

(**36.**) $(3b-x)(b-c) = 4c(b+x)$. $\therefore x = \frac{7bc-3b^2}{-b-3c}$, or $\frac{b(7c-3b)}{-b-3c}$.

(**37.**) $\frac{3bx}{2a^2} - \frac{x-b}{a+b} - \frac{bx-a^2}{a^2-b^2} + \frac{x}{4a} = 0$. $\therefore x = \frac{4a^2(ab-b^2+a^2)}{3a^3-6a^2b+ab^2+6b^3}$.

(**38.**) $\frac{3abc}{a+b} + \frac{a^2b^2}{(a+b)^3} + \frac{(2a+b)b^2x}{a(a+b)^2} = 3cx + \frac{bx}{a}$. $\therefore x = \frac{ab}{a+b}$.

**(39.)** $\frac{x-a}{b} + \frac{x-b}{c} + \frac{x-c}{a} = \frac{x-(a+b+c)}{abc}$. $\therefore x = \frac{a^2c + b^2a + c^2b - a - b - c}{ac + bc + ab - 1}$.

**(40.)** $\frac{7x+16}{21} - \frac{x+8}{4x-11} = \frac{x}{3}$. First, multiply by 21, and reduce; whence $\frac{21x+168}{4x-11} = 16$. $\therefore x = 8$.

**(41.)** $\frac{4x+13}{2x+3} + 1 = \frac{9x+22}{3x+4}$. $\therefore \frac{7}{2x+3} = \frac{10}{3x+4}$. $\therefore x = 2$.

**(42.)** $\frac{\frac{2}{5}x - 5}{2x-6} - \frac{x-2}{5x-15} = \frac{12x-7}{6x^2-54}$. $\therefore x = -\frac{154}{123}$.

**(43.)** $\frac{2x}{2-x} - \frac{2+6x}{2-3x} = 2 - \frac{1-2x}{2-x}$. $\therefore x = 3\frac{1}{3}$.

**(44.)** $\frac{x-3a}{x} = 5 - \frac{6}{x}$. $\therefore x = \frac{6-3a}{4}$.

**(45.)** $\frac{1}{x} = bc + d + \frac{ab}{x}$. $\therefore x = \frac{1-ab}{bc+d}$.

**(46.)** $\frac{1}{2}(\frac{2}{3}x + 4) - \frac{7\frac{1}{2} - x}{3} = \frac{x}{2}\left(\frac{6}{x} - 1\right)$. $\therefore x = 3$.

**(47.)** $\frac{1}{ab-ax} + \frac{1}{bc-bx} = \frac{1}{ac-ax}$. $\therefore x = \frac{b(a-b+c)}{a}$.

**(48.)** $\frac{x-7}{x+7} + \frac{1}{2(x+7)} = \frac{2x-15}{2x-6}$. $\therefore x = 8$.

**(49.)** $\frac{10x+3}{5x-4} + 1 = \frac{418}{25x^2-16} + \frac{15x-7}{5x+4}$. $\therefore x = 3$

**(50.)** $\frac{1}{n} + \frac{n}{n+x} = \frac{n+x}{nx}$. $\therefore x = \frac{n}{n-1}$.

**(51.)** $n + \frac{n-x}{n} = \frac{x^2}{n(n-x)}$. $\therefore x = \frac{n(n+1)}{n+2}$

**(52.)** $\frac{6x+a}{4x+b} = \frac{3x-b}{2x-a}$. $\therefore x = \frac{a^2-b^2}{b-4a}$.

**(53.)** $\frac{ad}{(c+dx)d} + \frac{b}{d} = \frac{2a-bx}{c+dx}$. $\therefore x = \frac{ad-bc}{2bd}$.

**(54.)** $ax^2 + a^3 = (ax + b^2)(a + x)$. $\therefore x = \frac{a(a^2 - b^2)}{a^2 + b^2}$.

**(55.)** $\frac{2a^2}{a - x} + \frac{5ax - x^2}{a - x} + \frac{x(a + x)}{a - x} = \frac{a}{b} - \frac{ac}{b(b + c)}$. $\therefore x = \frac{a(1 - 2b - 2c)}{1 + 6b + 6c}$.

**(56.)** $\frac{18x - 19}{28} + \frac{11x + 21}{6x + 14} = \frac{9x + 15}{14}$. $\therefore x = 7$.

**(57.)** $\frac{2x + 8\frac{1}{2}}{9} - \frac{13x - 2}{17x - 32} + \frac{x}{3} = \frac{7x}{12} - \frac{x + 16}{36}$. $\therefore x = 4$.

**(58.)** $\frac{7x + 6}{28} - \frac{2x + 4\frac{2}{7}}{23x - 6} + \frac{x}{4} = \frac{11x}{21} - \frac{x - 3}{42}$. $\therefore x = 4$.

**(59.)** $\frac{a^2(a + c)}{(a - b)(x - a)} - \frac{a^2(b + c)}{(a - b)(x - b)} = \frac{(a^3 + a^2x)n}{x^2 - a^2}$. $\therefore x = \frac{bn + c}{n - 1}$.

**(60.)** $\frac{3 + 2x}{2 - x} + \frac{a^3 + a^2b}{a^2b - b^3} + \frac{16x - x^2}{x^2 - 4} = \frac{2 - 3x}{2 + 3x} + \frac{a + b}{b} + \frac{2a}{a - b}$. $\therefore x = \frac{a - 3b}{b}$.

**(61.)** $\frac{(x - 1)(x + 1)}{3a} = \frac{2x + 2}{3a}$. $\therefore x = 3$.

**(62.)** $(x + 1)^2 = \{6 - (1 - x)\}x - 2$. $\therefore x = 1$.

**(63.)** $\frac{1}{x - 2} - \frac{1}{x - 4} = \frac{1}{x - 6} - \frac{1}{x - 8}$. $\therefore x = 5$.

**(64.)** $\left(\frac{x - a}{x + b}\right)^3 = \frac{x - 2a - b}{x + a + 2b}$. $\therefore x = \frac{a - b}{2}$.

**(65.)** $\frac{cx^m}{a + bx} = \frac{dx^m}{e + fx}$. $\therefore x = \frac{ad - ce}{cf - bd}$.

**(66.)** $\frac{a(b^2 + x^2)}{bx} = ac + \frac{ax}{b}$. $\therefore x = \frac{b}{c}$.

## SIMPLE EQUATIONS CONTAINING RADICALS.

(PAGE 231, *26.*)

**(1.)** $\sqrt{3y+3}=6. \quad \therefore y=11.$

**(2.)** $\sqrt{2z-5}=7. \quad \therefore z=27.$

**(3.)** $\sqrt{11x-2}=8. \quad \therefore x=6.$

**(4.)** $\sqrt{13x+39}-7=6. \quad \therefore x=10.$

**(5.)** $12+\sqrt{7y-3}=21. \quad \therefore y=12.$

**(6.)** $7\sqrt{5x+1}-12=30. \quad \therefore x=7.$

**(7.)** $b+\sqrt{mx-ac}=n. \quad \therefore x=\dfrac{(n-b)^2+ac}{m}.$

**(8.)** $\frac{1}{5}=-y+\sqrt{3y+y^2}. \quad \therefore y=\frac{1}{65}.$

**(9.)** $my+mn+m\sqrt{2my+y^2}=0. \quad \therefore y=\dfrac{n^2}{2(m-n)}.$

**(10.)** $\sqrt{x+40}-6=4-\sqrt{x}. \quad \therefore x=9.$

**(11.)** $2+\sqrt{x-24}=\sqrt{x}. \quad \therefore x=49.$

**(12.)** $2\sqrt{a^2+x^2}=4(a-\frac{1}{2}x). \quad \therefore x=\dfrac{3a}{4}.$

**(13.)** $\dfrac{\sqrt{a^2-x^2}}{\sqrt{a-x}}+x=a+2x. \quad \therefore \sqrt{a+x}=a+x. \quad \therefore x=1-a.$

**(14.)** $\sqrt{2+x}+\sqrt{x}=\dfrac{4}{\sqrt{2+x}}. \quad \therefore x=\frac{2}{3}.$

**(15.)** $\sqrt{x}+\sqrt{a-\sqrt{ax+x^2}}=\sqrt{a}. \quad \therefore x=\frac{9}{16}a.$

**(16.)** $x+a+\sqrt{2ax+x^2}=b. \quad \therefore x=\dfrac{(b-a)^2}{2b}$

**(17.)** $2\sqrt{a+x}-\sqrt{a-x}=\sqrt{a-x+\sqrt{ax+x^2}}. \quad \therefore x=\dfrac{64a}{1025}.$

**(18.)** $\sqrt{1+x}+\sqrt{1+x+\sqrt{1-x}}=\sqrt{1-x}, \quad \therefore x=-\dfrac{24}{25}.$

**(19.)** $\sqrt{x} + \sqrt{a + x} = \dfrac{na}{\sqrt{a + x}}$. $\therefore x = \dfrac{a(n - 1)^2}{2n - 1}$.

**(20.)** $\dfrac{\sqrt{ax} + \sqrt{b}}{\sqrt{ax} - \sqrt{b}} = \dfrac{\sqrt{a} + \sqrt{b}}{\sqrt{b}}$. $\therefore x = \dfrac{b(\sqrt{a} + 2\sqrt{b})^2}{a^2}$

**(21.)** $\dfrac{3\sqrt{x} - 4}{2 + \sqrt{x}} = \dfrac{15 + \sqrt{9x}}{40 + \sqrt{x}}$. $\therefore x = 4$.

**(22.)** $\dfrac{5x - 9}{\sqrt{5x} + 3} - 1 = \dfrac{\sqrt{5x} - 3}{2}$. $\therefore x = 5$.

**(23.)** $\dfrac{x - 1}{1 + \sqrt{x}} = 4 + \dfrac{\sqrt{x} - 1}{2}$. $\therefore x = 81$.

**(24.)** $\dfrac{3\sqrt{x} - 4}{2 + \sqrt{x}} = \dfrac{15 + \sqrt{9x}}{40 + \sqrt{x}}$. $\therefore x = 4$.

**(25.)** $\dfrac{\sqrt{a} + \sqrt{a - x}}{\sqrt{a} - \sqrt{a - x}} = \dfrac{1}{a}$. $\therefore x = \dfrac{4a^2}{(1 + a)^2}$.

**(26.)** $\dfrac{x}{\sqrt{b^2 + x} + b} = c$. $\therefore x = c^2 + 2bc$.

**(28.)** $\dfrac{24\sqrt{x}}{3 - 5x} = \dfrac{5\sqrt{x}}{7 - 3x}$. $\therefore x = 3\frac{12}{47}$.

**(29.)** $\sqrt[3]{a^2 + c} = \sqrt[4]{\dfrac{a^2 + c}{d(x + 9)}}$. $\therefore x = \dfrac{1}{d\sqrt[3]{a^2 + c}} - 9$.

**(30.)** $(x^{\frac{1}{2}} + 3a^2)^{\frac{1}{2}} - (x^{\frac{1}{2}} - 3a^2)^{\frac{1}{2}} = \dfrac{2a^{\frac{1}{2}}x^{\frac{1}{2}}}{b^{\frac{1}{2}}}$. $x = \dfrac{9a^3b^2}{4(b - a)}$.

**(31.)** $\sqrt[m]{a + x} = \sqrt[2m]{x^2 + 5ax + b^2}$. $\therefore x = \dfrac{a^2 - b^2}{3a}$.

**(32.)** $\sqrt{(1 - a)^2 + (1 + a)x} - \sqrt{(1 + a)^2 + (1 - a)x} = 2a$. $\therefore x = 8$.

**(33.)** $8.4x - 7.6 = 10 + 2.2x$. $\therefore x = 2.838$.

**(34.)** $3.75x + .5 = 2.25x + 8$. $\therefore x = 5$.

**(35.)** $.2x + 6.1 = \dfrac{2x + 5}{4} + 100x + 1.5401$. $\therefore x = .033$.

## APPLICATIONS OF SIMPLE EQUATIONS.

### (Page 238, *32.*)

(**1.**) What number is that to $\frac{1}{5}$ of which, if 17 be added, the sum is 22? *Ans.*, 25.

Stoddard's Intellectual Arithmetic, pp. 116—140, affords so ample a collection of excellent examples under this head, that it is not thought necessary to give more here. A few illustrations of the literal notation are given. They are not numbered with reference to the text-book.

(**1.**) A and B bought an equal number of horses; but afterward A sold $c$ horses, and B bought $d$, when they together had $m$ horses. How many did each buy? *Ans.*, Each bought $\frac{m - d + c}{2}$ horses.

(**2.**) A man bought a sheep, a cow, and a horse, for \$$s$. The cow cost \$$a$ more than the sheep, and \$$b$ less than the horse. What was the cost of each? Apply this result to example 20, p. 117, in the Int. Ar.

*Ans.*, A sheep cost $\frac{s - 2a - b}{3}$; a cow, $\frac{s + a - b}{3}$; and a horse, $\frac{s + a + 2b}{3}$.

(**3.**) A farmer bought a plow, a harness, and a horse, for \$$s$. For the harness he gave \$$a$ more than for the plow, and \$$b$ less than for the horse. What was the price of each? How does this example differ from the preceding? (The answers are the same in form.)

(**4.**) An $\frac{m}{n}$th part of a certain number minus $a$, is $s$. What is the number? *Ans.*, $(s + a)\frac{n}{m}$. Apply this result to examples 25, 26, in the Int. Ar., p. 117.

(**5.**) A man bought a watch and chain for \$$s$. The chain cost $\frac{m}{n}$th part as much as the watch. What was the cost of the watch?

*Ans.*, \$$\frac{ns}{m + n}$. Apply this, as above, to example 22, p. 117.

(**6.**) Certain supplies would last a garrison $a$ days; but if $b$ men were removed the supplies would last the remainder $c$ days. How many days supply for one man was there? How many men were there in the garrison at first?

*Ans.*, Day's supply, $\frac{abc}{c-a}$; men at first, $\frac{bc}{c-a}$.

(**7.**) When my father was $a$ years old my age was $\frac{1}{m}$th of his. What will his age be when mine is $\frac{1}{n}$th of his? *Ans.*, $\frac{an(m-1)}{m(n-1)}$.

(**8.**) A can do a piece of work in $a$ days, and A and B together, in $b$ days. After A has done $\frac{1}{n}$th part of it, how long will it take B to finish it? *Ans.*, $\frac{ab(n-1)}{n(a-b)}$.

(**9.**) A, B, and C, talking of their ages, A says to B, "I am $a$ times as old as you;" B says to C, "I am $\frac{1}{b}$th as old as you;" but A says to C, "I am $m$ years older than you." What was the age of each?

*Ans.*, A's age is $\frac{am}{a-b}$; B's, $\frac{m}{a-b}$; and C's, $\frac{bm}{a-b}$.

(**10.**) A man agreed to work $s$ days on these conditions: for every day he worked he was to receive \$$a$ and for every day he was idle he was to forfeit \$$b$. At the end of the time he received \$$m$. How many days did he work, and how many was he idle?

*Ans.*, He worked $\frac{m+bs}{a+b}$ days, and was idle $\frac{as-m}{a+b}$ days.

Apply the above results to examples 1, 2, 3, page 123, of the Int. Ar.

(**11.**) A farmer wishing to buy a certain number of sheep, found that if he gave \$$a$ a head, he would have \$$m$ remaining; but, if he gave \$$b$ a head, he would lack \$$n$ of having money enough. How many sheep did he wish to buy; and how much money had he?

*Ans.*, $\frac{m+n}{b-a}$ sheep, and \$$\frac{an+bm}{b-a}$.

(**12.**) A general lost in battle an $\frac{m}{n}$th part of his army, killed, and an $\frac{a}{b}$th part taken prisoners. He then had $s$ men left. How many had he at first? *Ans.*, $\frac{bns}{b(n-m)-an}$.

(**13.**) There is a fish which weighs $s$ lbs.; and $\frac{1}{n}$th of the weight of the head $+ m$ lbs., equals the weight of the body. What is the weight of each part separately?

*Ans.*, The head, $\frac{n(s-m)}{n+1}$; the body, $\frac{s+mn}{n+1}$.

(**14.**) The head of a fish is $a$ inches long; its tail is $\frac{1}{n}$th of the length of its body $+ b$ inches; and its body is $m$ inches longer than its head and tail together. What is the length of the fish?

*Ans.*, $\frac{2n(a+b)+m(n+1)}{(n-1)}$.

(**15.**) If one of two numbers be multiplied by $m$, and the other by $n$, the sum of the products is $p$; but if the first be multiplied by $m'$, and the second by $n'$, the product is $p'$. What are the numbers?

*Ans.*, $\frac{n'p-np'}{mn'-m'n}$, $\frac{mp'-m'p}{mn'-m'n}$.

(**16.**) An ingot of metal which weighs $n$ pounds loses $p$ pounds when weighed in water. This ingot is composed of two metals $M$, and $M'$. $n$ pounds of $M$ loses $q$ pounds when weighed in water; and $n$ pounds of $M'$ loses $r$ pounds. How much of each metal does the ingot contain? *Ans.*, $\frac{n(r-p)}{r-q}$ pounds of $M$, and $\frac{n(p-q)}{r-q}$ of $M'$.

## TRANSLATION OF EQUATIONS INTO PRACTICAL PROBLEMS.

(PAGE 250, ***35.***)

Examples of this class can be formed from any of the examples in the text-book or in this collection from ***20*** to ***26.***

## ELIMINATION BY COMPARISON.

(PAGE 255, ***41.***)

(**1.**) $2x+3y=23$, and $5x-2y=10$. $\therefore$ $x=4$, $y=5$.

(**2.**) $6x-2y=14$, and $5x-6y=-10$. $\therefore$ $x=4$, $y=5$.

(**3.**) $11x+3y=100$, and $4x-7y=4$. $\therefore$ $x=8$, $y=4$.

(4.) $\frac{x}{7} + 7y = 99$, and $\frac{y}{7} + 7x = 51$. $\therefore\ x = 7,\ y = 14$.

(5.) $\frac{x+2}{3} + 8y = 31$, and $\frac{y+5}{4} + 10x = 192$. $\therefore\ x = 19,\ y = 3$.

(6.) $x + y = a,\ bx + cy = de$. $\therefore\ x = \frac{ac - de}{c - b},\ y = \frac{de - ab}{c - b}$.

## ELIMINATION BY SUBSTITUTION.

(Page 257, ***42.***)

The preceding examples and the corresponding ones in the text-book will afford good and sufficient exercise, under this topic.

## ELIMINATION BY ADDITION OR SUBTRACTION.

Assign the examples in either or both of the two preceding sections, and require that the elimination be performed by this method.

## GENERAL EXERCISES.

(Pages 263—265.)

(1.) $2x - 3y = 1$, and $5x + 2y = 31$. $\therefore\ x = 5,\ y = 3$.

(2.) $\frac{8x}{7} = \frac{5(x-y)}{3} - 6$, and $\frac{3x-y}{9} + 5 = \frac{4y+3x}{6}$. $\therefore\ x = 18\frac{36}{37}$, $y = \frac{612}{259}$.

(3.) $5x - 8\frac{1}{2} = 7y - 44$, and $2x = y + \frac{5}{7}$. $\therefore\ x = 4\frac{1}{2},\ y = 8\frac{2}{7}$.

(4.) $2x - \frac{y+3}{4} = 8$, and $4y - \frac{8-x}{3} = 24\frac{1}{2} - \frac{2y+1}{2}$. $\therefore\ x = 5$, $y = 5$.

(5.) $\frac{4x}{x^2} + \frac{5y}{y^2} = \frac{9}{y} - 1$, and $\frac{5}{x} + \frac{4}{y} = \frac{7}{x} + \frac{3}{2}$. $\therefore\ x = 4,\ y = 2$.

(6.) $\frac{a}{x} + \frac{b}{y} = m$, and $\frac{c}{x} + \frac{d}{y} = n$. $\therefore\ x = \frac{bc - ad}{bn - dm},\ y = \frac{bc - ad}{cm - an}$.

(7.) $\frac{2r}{3}+\frac{4s}{5}=64$, and $\frac{5r}{6}+\frac{9s}{10}=77$. $\therefore\ r=60,\ s=30$.

(8.) $5+\frac{y-8}{5}=\frac{3x+4y+3}{10}-\frac{2x+7-y}{15}$, and $\frac{7x+6}{11}=\frac{9y+5x-8}{12}-\frac{x+y}{4}$. $\Big\}$ $x=7$, $y=9$.

(9.) $\dfrac{\frac{2x}{3}-\frac{5y}{12}}{\frac{7}{4}}-\dfrac{\frac{3x}{2}-\frac{y}{3}}{\frac{23}{2}}=2$, and $\frac{x-y}{x+y}=\frac{1}{5}$. $\therefore\ x=18,\ y=12$.

(10.) $x-\frac{3y-2+x}{11}=1+\frac{15x+\frac{4}{3}y}{33}$, and $\frac{3x+2y}{6}-\frac{y-5}{4}=\frac{11x+152}{12}-\frac{3y+1}{2}$. $\Big\}$ $x=8$, $y=9$.

(11.) $\frac{21}{14+x}=\frac{14}{84+y}$, and $21x+28y=334$. $\therefore\ x=61\frac{15}{119}$, $y=-33\frac{108}{119}$.

(12.) $4x+3y+\frac{24+5\frac{1}{2}y}{2x+1}=\frac{16x^2+12xy-8x+5y+28}{4x-2}$, and $2x+4=3y+\frac{8x^2-18y^2+108}{4x+6y+3}$. $\Big\}$ $x=3$, $y=2$.

(13.) $\frac{a}{b+y}=\frac{b}{3a+x}$, and $ax+2by=c$. $\therefore\ x=\frac{2b^2-6a^2+c}{3a}$, $y=\frac{3a^2-b^2+c}{3b}$.

(14.) $y=x+18.73$, and $-0.56x+13.421y=763.4$. $\therefore\ x=39.8121+$, $y=58.5421+$.

(15.) $1.2x+3.6y=\frac{2-x}{3}-.5y$, and $x-.3y=\frac{1-.4y}{5}$. $\Big\}$ $x=.217+$, $y=.0811+$.

---

## APPLICATIONS.

### (Pages 265—271.)

(1.) A and B engage in play; in the first game A wins as much as he had and \$4 more; in the second game B wins $\frac{1}{2}$ as much as he had at first and \$1 more, when it appears that he has 3 times as much as A. What had each at first? *Ans.*, A, \$6; B, \$18.

(**2.**) A sum of money was divided equally among a certain number of persons; had there been 3 more, each would have received \$1 less, and had there been 2 fewer, each would have received \$1 more than he did. How many persons were there, and what did each receive?

*Ans.*, 12 persons, and \$5.

(**3.**) A certain fraction becomes 1 when 3 is added to its numerator, and $\frac{1}{2}$ when 2 is added to its denominator. What is it? *Ans.*, $\frac{5}{8}$.

(**4.**) Find two numbers, such that $\frac{1}{2}$ the first and $\frac{3}{4}$ of the second together may be equal to the difference of 3 times the first and second, and this difference equal to 11. *Ans.*, 7, and 10.

(**5.**) What fraction is that which becomes $\frac{1}{2}$ when numerator and denominator are each diminished by 1, and $\frac{1}{2}$ when its denominator alone is increased by 1?

*Ans.*, $\frac{6}{11}$; or any other fraction whose numerator doubled equals its denominator $+$ 1.

(**6.**) What fraction is that which becomes $\frac{1}{2}$ when its numerator is increased by 1 and its denominator diminished by 1; but which becomes $\frac{1}{3}$ when its numerator is doubled and its denominator increased by 5? *Ans.*, $\frac{2}{7}$.

(**7.**) Two men, A and B, received \$117 for their wages; A having been employed 15 weeks, and B 14 weeks. A received for 4 weeks \$11 more than B did for 3 weeks. What were their weekly wages?

*Ans.*, A's, \$5; B's, \$3.

(**8.**) After A had won \$4 from B, he had only half as many dollars as B had left; and had B won \$6 from A, he would have 3 times as many as A had left. How many had each? *Ans.*, A, \$36; B, \$84.

(**9.**) Divide 50 into two such parts, that $\frac{3}{4}$ of one part, added to $\frac{5}{6}$ of the other, will be equal to 40. *Ans.*, 20 and 30.

(**10.**) Find two numbers, whose difference is 8, and the difference of their squares 208. *Ans.*, 9 and 17.

(**11.**) A and B can perform a piece of work in 16 days. After working together 4 days, B can finish it in 36 days more. In what time would each do it alone? *Ans.*, A in 24, and B in 48.

(**12.**) A and B, jointly, have a fortune of \$9800. A invests $\frac{1}{6}$ of his fortune, and B $\frac{1}{5}$ of his, in trade, and still each has the same sum remaining. Required the fortune of each. *Ans.*, A's, \$4800; B's, \$5000.

(**13.**) There is a certain number, consisting of two digits. The *sum* of those digits is 5; and if 9 be added to the number itself, the digits will be inverted. What is the number? *Ans.*, 23.

(**14.**) There is a number consisting of two digits, the second of which is greater than the first, and if the number be divided by the sum of its digits, the quotient is 4; but if the digits be inverted, and that number divided by a number greater by 2 than the difference of the digits, the quotient becomes 14. Required the number.

*Ans.*, 48.

(**15.**) Find two numbers such that if 5 be divided by the first, and 11 by the second, the sum of the quotients shall be $7\frac{1}{6}$; but if 5 be divided by the second, and 11 by the first, the sum of these quotients will be 1 less than in the former case. *Numbers*, 3, and 2.

(**16.**) Find two numbers such that the product of the first increased by 5, into the second increased by 7, shall exceed the product of the first increased by 1, into the second diminished by 9, by 112; while twice the first + 10 equals 3 times the second + 1.

*Numbers*, 3, and 5.

(**17.**) Find *formulæ* for the values of two numbers such that if $a$ be divided by the second $+ b$, the quotient will equal $b$ divided by the first increased by $3a$; while $a$ times the first $+ 2b$ times the second equals $c$.

$$\textit{Formulæ},\ x = \frac{2b^2 - 6a^2 + c}{3a},\ y = \frac{3a^2 - b^2 + c}{3b}.$$

(**18.**) A railway train after running for one hour meets with an accident which delays it one hour, after which it proceeds at $\frac{3}{5}$ of its former rate, and arrives at the terminus 3 hours behind time; had the accident occurred 50 miles further on, the train would have arrived 1 hour and 20 minutes sooner. Required the length of the line and the first rate of speed. *Ans.*, 100 miles, and 25.

SUG.— Let $x$ = the distance and $y$ the first rate. Then $x - y$ is the distance run after the accident and at a rate of $\frac{3}{5}y$ miles per hour. Then $\frac{x-y}{\frac{3}{5}y}$ (the running time after the accident) + 1 (the hour's delay) + 1 (the time run before the accident) $= \frac{x}{y}$ (the regular running time for the whole distance) + 3 (the additional time for detention). (1) $\frac{x-y}{\frac{3}{5}y} + 2 = \frac{x}{y} + 3$. In a similar manner $\frac{x-y-50}{\frac{3}{5}y} + 2 + \frac{50}{y} = \frac{x}{y} + 1\frac{2}{3}$ is found to be the second equation.

(**19.**) A banker has two kinds of money; it takes $a$ pieces of the first to make a dollar, and $b$ pieces of the second to make the same sum. If he is offered a dollar for $c$ pieces, how many of each kind must he give?

*Ans.*, First kind, $\frac{a(c-b)}{a-b}$; second kind, $\frac{b(a-c)}{a-b}$.

In the last example, if $a = 10$, $b = 20$, and $c = 15$, how many of each kind must he give? *Ans.*, First kind, 5; second kind, 10.

(**20.**) A vinter has two casks of wine, from the greater of which he draws 15 gallons, and from the less, 11; he then finds that the former contains $2\frac{2}{3}$ times as much as the latter. After they became half empty he put 10 gallons of water into each, and then found the former contained $1\frac{4}{5}$ times as much of the mixture as the latter. How many gallons did each cask hold? *Ans.*, 79, and 35.

(**21.**) A person laid out a rectangular garden, and found that if each side had been 4 yards longer, the longer side would have been $1\frac{1}{4}$ times the shorter; but, if each side had been 4 yards shorter, the longer would have been $1\frac{1}{3}$ times the shorter. What were the lengths of the sides? *Ans.*, 36 and 28.

(**22.**) When wheat was 5 shillings a bushel, and rye 3 shillings, a man wanted to fill his sack with a mixture of rye and wheat for the money he had in his purse. If he bought 7 bushels of rye and laid out the rest of his money in wheat, he would want 2 bushels to fill his sack; but if he bought 6 bushels of wheat, and filled his sack with rye, he would have 6 shillings left. How must he lay out his money, and fill his sack?

*Ans.*, He must buy 9 bushels of wheat and 12 bushels of rye.

(**23.**) A composition of copper and tin, containing 100 cubic inches, weighs 505 ounces. How many ounces of each metal does it contain, supposing the weight of a cubic inch of copper to be $5\frac{1}{4}$ ounces, and of a cubic inch of tin $4\frac{1}{4}$ ounces? *Ans.*, 420, and 85 ounces.

(**24.**) A work is to be printed so that each page may contain a certain number of lines, and each line a certain number of letters. If each page should contain 3 lines more, and each line 4 letters more, then there would be 224 letters more on each page; but if there should be 2 lines less on a page, and 3 letters less in each line, then each page would contain 145 letters less. How many lines are there on each page, and how many letters in each line?

*Ans.*, 29 lines on a page, and 32 letters in a line.

## QUATIONS WITH MORE THAN TWO UNKNOWN QUANTITIES.

(PAGE 272, ***46.***)

**(1.)**
$$\left.\begin{array}{r} x - y + z = 30 \\ 8x - 4y + 2z = 50 \\ 27x - 9y + 3z = 64 \end{array}\right\} \therefore \left\{\begin{array}{l} x = \frac{2}{3}, \\ y = 7, \\ z = 36\frac{1}{3}. \end{array}\right.$$

**(2.)**
$$\left.\begin{array}{r} 2x + 2y = 20 \\ 3x + 3z = 57 \\ 4y + 4z = 92 \end{array}\right\} \therefore \left\{\begin{array}{l} x = 3, \\ y = 7, \\ z = 16. \end{array}\right.$$

**(3.)**
$$\left.\begin{array}{l} 2x + 5y + 7z = -288 \\ 5x - y - 3z = 227 \\ 7x + 6y - z = 297 \end{array}\right\} \therefore \left\{\begin{array}{l} x = 13, \\ y = 24, \\ z = -62. \end{array}\right.$$

**(4.)**
$$\left.\begin{array}{l} y + \frac{1}{2}x = 41 \\ x + \frac{1}{4}z = 20\frac{1}{2} \\ y + \frac{1}{5}z = 34 \end{array}\right\} \therefore \left\{\begin{array}{l} x = 18, \\ y = 32, \\ z = 10. \end{array}\right.$$

**(5.)**
$$\left.\begin{array}{r} ax = c + by \\ f - ey = dx \\ l + gy = hz \end{array}\right\} \therefore \left\{\begin{array}{l} x = \dfrac{ce + bf}{ae + bd}, \\ y = \dfrac{af - cd}{ae + bd}, \\ z = \dfrac{a(el + fg) + d(bl - cg)}{h(ae + bd)}. \end{array}\right.$$

**(6.)**
$$\left.\begin{array}{l} 0 = y + \frac{1}{2}x - \frac{1}{2}z - 62 \\ 26 - \frac{1}{8}y = \frac{1}{4}x \\ 5y - 4z = 0 \end{array}\right\} \therefore \left\{\begin{array}{l} x = 64, \\ y = 80, \\ z = 100. \end{array}\right.$$

**(7.)**
$$\left.\begin{array}{r} 18x = -11 + 7y + 5z \\ 114\frac{2}{3} + z = \frac{2}{3}x - 4\frac{2}{3}y \\ 3\frac{1}{2}z + 80 = -2y - \frac{3}{4}x \end{array}\right\} \therefore \left\{\begin{array}{l} x = -12, \\ y = -25, \\ z = -6. \end{array}\right.$$

**(8.)**
$$\left.\begin{array}{r} 2x - y + 3z - v + 5t = 8 \\ -x + y - z + 4v - t = 6 \\ -3x + 4y - z + 6v - 2t = 5 \\ 5x + y + z - 2v + 3t = 4 \\ x - 3y + 3z - 2v + 4t = 1\frac{1}{2} \end{array}\right\} \therefore \left\{\begin{array}{l} x = -\frac{132}{102}, \\ y = \frac{21}{102}, \\ z = -\frac{672}{102}, \\ v = \frac{108}{62}, \\ t = \frac{545}{102}. \end{array}\right.$$

(**9.**) $$\left.\begin{aligned} x + 2z + 3v + 4t &= 14 \\ -x + 3y + 2z + 5v + 3t &= 64\tfrac{1}{2} \\ x + 6y - z + 8v &= 70 \\ -x + y - z + 4v - t &= 22\tfrac{1}{2} \\ x - 3y + 3z - 2v + 4t &= -24 \end{aligned}\right\} \therefore \left\{\begin{aligned} x &= -7, \\ y &= 10, \\ z &= 7, \\ v &= 3, \\ t &= -\tfrac{1}{2}. \end{aligned}\right.$$

(**10.**) $\dfrac{2x - 3y + 8z}{u + 1} = 1$, $\dfrac{3y - 2z}{4x} = \dfrac{7}{8}$, $\dfrac{4u - 11}{4z - 5} = 3$, and $\dfrac{1}{x} + \dfrac{1}{y} = \dfrac{5}{2y}$.
$\therefore\ x = 4$, $y = 6$, $z = 2$, and $u = 5$.

(**11.**) $$\left.\begin{aligned} ax + by + cz &= 3 \\ ax + by - cz &= 1 \\ ax - by + cz &= 1 \end{aligned}\right\} \therefore \left\{\begin{aligned} x &= \frac{1}{a}, \\ y &= \frac{1}{b}, \\ z &= \frac{1}{c}. \end{aligned}\right.$$

(**12.**) $x + y = axy$, $x + z = bxz$, and $y + z = cyz$. $\therefore\ x = \dfrac{2}{a + b - c}$, $y = \dfrac{2}{a - b + c}$, and $z = \dfrac{2}{b + c - a}$.

SUG.—From these we readily obtain $\dfrac{1}{y} + \dfrac{1}{x} = a$, $\dfrac{1}{z} + \dfrac{1}{x} = b$, and $\dfrac{1}{z} + \dfrac{1}{y} = c$.

(**13.**) $$\left.\begin{aligned} 2yz - \tfrac{5}{3}xz + xy &= \tfrac{85}{27}xyz \\ \tfrac{1}{4}yz + xz + 2xy &= \tfrac{443}{72}xyz \\ \tfrac{5}{6}yz - xz + 4xy &= \tfrac{433}{36}xyz \end{aligned}\right\} \therefore \left\{\begin{aligned} x &= 6, \\ y &= 9, \\ z &= \tfrac{1}{3}. \end{aligned}\right.$$

---

## APPLICATIONS.

### (PAGES 277—279.)

(**1.**) Divide the number $a$ into three such parts that the first and second divided by the third give, respectively, $m$ and $n$.

*Ans.*, $\dfrac{a}{m + n + 1}$, $\dfrac{ma}{m + n + 1}$, and $\dfrac{na}{m + n + 1}$.

(**2.**) The sums of each pair of three numbers are respectively $m$, $n$, and $p$; find them. *Ans.*, $\dfrac{m + n - p}{2}$, $\dfrac{m - n + p}{2}$, $\dfrac{-m + n + p}{2}$.

(**3.**) There is a number consisting of 3 digits, such, that if 198 be added to it, the digits will be inverted. The sum of the digits is 18. Also, 5 times the units' digit is equal to 7 times the digit in the place of hundredths. Required the number. *Ans.*, 567.

(**4.**) Two pipes, A and B, will fill a cistern in 70 minutes, A and C will fill it in 84 minutes, and B and C in 140 minutes. How long will it take each to fill the cistern alone?

*Ans.*, A in 105 m.; B in 210 m.; and C in 420 m.

(**5.**) Find 4 numbers, such that the 1st plus $\frac{1}{2}$ the 2d $= 357$; the 2nd plus $\frac{1}{3}$ the 3rd $= 476$; the 3rd plus $\frac{1}{4}$ the 4th $= 595$; and the 4th plus $\frac{1}{5}$ the 1st $= 714$. *Ans.*, 190, 334, 426, and 676.

(**6.**) A worked 10 days, B 4 days, and C 3 days, and their wages amounted to $23; at another time, A worked 9 days, B 8 days, and C 6 days, and their wages amounted to $24; a third time, A worked 7 days, B 6 days, and C 4 days, and their wages amounted to $18. What were the daily wages of each?

*Ans.*, A's, $2.00; B's, 0; and C's, $1.00.

(**7.**) Find what each of three persons, A, B, and C, is worth, from knowing, 1st, that what A is worth added to 3 times what B and C are worth, make 4700 dollars; 2nd, that what B is worth added to 4 times what A and C are worth make 5800 dollars; 3rd, that what C is worth added to 5 times what A and B are worth make 6300 dollars.

*Ans.*, A is worth 500, B 600, C 800 dollars.

(**8.**) Find what each of three persons, A, B, C, is worth, knowing, 1st, that what A is worth added to $l$ times what B and C are worth is equal to $p$; 2nd, that what B is worth added to $m$ times what A and C are worth is equal to $q$; 3rd, that what C is worth added to $n$ times what A and B are worth is equal to $r$.

*Ans.*, A is worth $\$\frac{(nq-r)(lm-l)-(lq-p)(mn-1}{(mn-n)(lm-l)-(lm-1)(mn-1)}$;

B " $\$\frac{(np-r)(lm-m)-(mp-q)(ln-1)}{(ln-n)(lm-m)-(lm-m)(ln-1)}$;

C " $\$\frac{(lq-p)(mn-n)-(nq-r)(lm-1)}{(lm-l)(mn-n)-(lm-1)(mn-1)}$.

(**9.**) A railway train meets with an accident which causes it to lessen its speed $\frac{1}{n}$th, and is in consequence $a$ hours late at the end of the line. If the accident had happened $b$ miles further on, the train would have been $c$ hours late. What was the first rate of running?

*Ans.*, $\frac{b(n-1)}{a-c}$ miles per hour

## RATIO.

(PAGE 283, **54.**)

**(1.)** $6xy^3 : 2xy = 3y^2$.

**(2.)** $(m-n)^2 : m^2 - n^2 = \frac{m-n}{m+n}$.

**(3.)** $\frac{3}{4} : \frac{5}{6} = \frac{9}{10}$.

$\frac{1}{7} : \frac{5}{14} = \frac{2}{5}$.

$(m+x)^2 : m^2 - x^2 = \frac{m+x}{m-x}$.

$\frac{a^2-b^2}{4mx} : \frac{a+b}{2x} = \frac{a-b}{2m}$.

$\frac{2c^2y^3}{5am^2} : \frac{6a^2y}{12cm^2} = \frac{4c^3y^2}{5a^3}$.

$\frac{(a-m)^3}{x^3+y^3} : \frac{a^2-m^2}{x+y} = \frac{(a-m)^2}{(x^2-xy+y^2)(a+m)}$.

**(4.)** The duplicate ratio of 6 to 4 is 9 : 4, or $\frac{9}{4}$, or $2\frac{1}{4}$.

**(5.)** The subtriplicate ratio of 27 to 125 is $\frac{3}{5}$.

**(6.)** The compound ratio of 2 : 5, 10 : 7, 21 : 16, and 4 : 3 is 1.

**(7.)** Reduce 187 : 55 to its lowest terms. *Result*, $\frac{17}{5}$.

**(8.)** $48 : 80 = \frac{3}{5}$. $a^2 - 2ax + x^2 : a^2 - x^2 = \frac{a-x}{a+x}$.

**(9.)** Show that $7 : 11 > 3 : 5$. $7 : 11 = \frac{35}{55}$, and $\frac{3}{5} = \frac{33}{55}$.

**(10.)** If the ratio of $a$ to $b$ is $\frac{3}{5}$, what is the ratio of $a+b$ to $b$, and of $b-a$ to $a$? *Ans.*, $1\frac{3}{5}$, and $\frac{2}{3}$.

**(11.)** If the ratio of $m$ to $n$ is $1\frac{3}{4}$, what is the ratio of $m-n$ to $6m$, and also to $5n$? *Ans.*, $\frac{1}{14}$, and $\frac{3}{20}$.

**(12.)** If the ratio of $m$ to $2m+3n$ is $\frac{5}{13}$, what is the ratio of $m$ to $n$? *Ans.*, 5 to 1.

**(13.)** If the ratio of $m$ to $n$ is $\frac{2}{7}$, what is the ratio of $12m$ to $m+n$, and of $12n$ to $n-2m$. *Ans.*, $2\frac{2}{3}$, and 28.

**(14.)** If the ratio of $5y-8x$ to $7x-5y$ is 6, what is the ratio of $x$ to $y$? *Ans.*, 7 to 10.

**(15.)** Which is greatest and which least of the ratios 2 : 3, 3 : 5, 6 : 11, 8 : 15, and 4 : 9?

*Ans.*, 2 : 3 is greatest; and 8 : 15 is least.

(**16.**) Is the ratio $a + 3x : a$ a ratio of greater or less inequality when $a$ and $x$ are positive? When $x$ is negative? When $x$ is 0? As $x$ approaches in value to $a$ toward what does the value of the ratio approach? As $x$ approaches 0 toward what does the value of the ratio approach? *Ans.*, Greater; Less; Unity; 4; and 1.

(**17.**) Compound the direct ratio $x^3 - y^3 : x^2 - y^2$, and the inverse ratio of $x - y$ to $x + y$. *Result*, $\frac{x^2 + xy + y^2}{x - y}$.

(**18.**) What is the value of the subduplicate ratio of $2ax^2$ to $3by^4$, compounded with the inverse triplicate ratio of $x$ to $y$, when $a = 2$, and $b = 3$? *Ans.*, $\frac{2y}{3x^2}$.

(**19.**) Which is the greater, the direct ratio of 3 to 4, or the inverse duplicate ratio of 4 to 3? *Ans.*, The first.

(**20.**) What quantity must be added to each of the terms of the ratio $m : n$, that it may become equal to $p : q$? *Ans.*, $\frac{mq - np}{p - q}$.

---

## PROPORTION.

### (Page 288, *69.*)

(**1.**) If $c : d :: a : b$, prove that $mc : md :: \frac{a}{n} : \frac{b}{n}$.

(**2.**) If $a : b :: c : d$, prove that $\frac{a}{m} : nb :: \frac{c}{m} : nd$.

(**3.**) If $m : n :: x : y$, is $am : n :: x : \frac{y}{n}$? Why?

(**4.**) If $a : b :: m : n$, is $a : 2b :: 2m : n$? Why?

(**5.**) If $a : b :: m : n$, is $3a : 2b :: 6m : 4n$? Why?

(**6.**) If $a : b :: x : y$, is $2a : 2b :: 2x : y$? Why?

If $a : b :: x : y$, is $2a : \frac{b}{3} :: 6x : y$? Why?

---

(**7 to 9.**) Prove that if $x : y :: m : n$, $y : n :: x : m$. Also that $n : m :: y : x$. Also that $m : x :: n : y$.

(**10.**) If $3m^2x : 5n^3y^2 :: 6mx : 10n^2y^2$, prove that $m = n$.

(**11.**) If $3a^2 : 4b :: \frac{a^3}{8} : \frac{1}{6}$, prove that $a = \frac{1}{b}$.

---

(**12.**) From $3 : 7 :: 9 : 21$ how is $10 : 3 :: 30 : 9$ obtained? How $12 : 3 :: 28 : 7$ ? How $4 : 7 :: 12 : 21$?

(**13.**) If $m : n :: x : y$ prove that $m \pm n : m :: x \pm y : y$.

(**14.**) If $m : n :: x : y$ prove that $m \pm x : x :: n \pm y : y$.

(**15.**) If $a : b :: m : n$ is $a + n : b + m :: a : b$?

(**16.**) If $6y : 4a :: 3y - 2x : 2a - \frac{1}{3}b$, show that $y : x :: 4a : b$.

(**17.**) If $a : b :: p : q$, show that $a^2 + b^2 : \frac{a^3}{a+b} :: p^2 + q^2 : \frac{p^3}{p+q}$.

SUG'S. $a^2 : b^2 :: p^2 : q^2$. $a^2 + b^2 : a^2 :: p^2 + q^2 : p^2$. Also $a + b : a :: p + q : q$. Now the two proportions, being equal ratios, may be multiplied together. Hence, $(a^2+b^2)(a+b) : a^3 :: (p^2+q^2)(p+q) : p^3$, and $a^2 + b^2 : \frac{a^3}{a+b} :: p^2 + q^2 : \frac{p^3}{p+q}$.

(**18.**) If $a : b :: x : y$, show that $a^{\frac{2}{3}} : b^{\frac{2}{3}} :: x^{\frac{2}{3}} : y^{\frac{2}{3}}$.

---

(**19.**) If $x$ and $y$ have the duplicate ratio of $x + z$ to $y + z$, show that $z$ is a mean proportional between $x$ and $y$.

SUG'S.—From $x : y :: (x + z)^2 : (y + z)^2$; $x^2y + 2xyz + yz^2 = xy^2 + 2xyz + xz^2$. $\therefore$ $x^2y - xy^2 = xz^2 - yz^2$; $xy(x - y) = z^2(x - y)$; $xy = z^2$. $\therefore$ $x : z :: z : y$.

(**20.**) What proportion is deducible from the equation $ab = a^2 - x^2$?

*Ans.*, $a : a + x :: a - x : b$.

(**21.**) What proportion is deducible from the equation $x^2 + y^2 = 2ax$?

*Ans.*, $x : y :: y : 2a - x$.

(**22.**) If $2a^2 = 1$, show that $a$ is to 1 inversely as $\sqrt{2}$ is to 1.

(**23.**) Prove that if the antecedent and consequent of a ratio be increased or diminished by like parts of each, the resulting quantities and the antecedent and consequent will be proportional.

SUG'S. $a \pm \frac{a}{m} : b \pm \frac{b}{m} :: a : b$, since $a \pm \frac{a}{m} : b \pm \frac{b}{m} = \frac{am \pm a}{m} : \frac{bm \pm b}{m} = a(m \pm 1) : b(m \pm 1) = a : b$.

(**24.**) If $a : b :: c : d$, $a : c :: m : n$, and $c : n :: x : m$, show that $b : d :: x : c$.

(**25.**) Show that if $a : x : y : b$, $x = \sqrt[3]{a^2b}$, and $y = \sqrt[3]{ab^2}$.

[NOTE. $a : x : y : b$, means $a : x :: x : y$, and $x : y :: y : b$. $x$ and $y$ are *two geometrical* means between $a$ and $b$.]

(**26.**) If $a : x : y : z : b$, show that $x = \sqrt[4]{a^3b}$, $y = \sqrt[4]{a^2b^2}$, and $z = \sqrt[4]{ab^3}$.

[NOTE.—The last two examples are but examples of inserting a number of means between two given extremes in a geometrical progression. See (***81***).]

---

## ARITHMETICAL PROGRESSION.

### (PAGES 297—299, ***74—76.***)

For additional examples see Complete Arithmetic (444, 445, 446, 447.)

(**1.**) Insert 15 *A. M.*, between 3 and 47.

*Ans.*, 3, $5\frac{3}{4}$, $8\frac{1}{2}$, etc., to $44\frac{1}{4}$, 47.

(**2.**) Insert 10 *A. M.*, between 3 and 58.

*Ans.*, 3, 8, 13, 18, 23, 28, 33, 38, 43, 48, 53, 58.

(**3.**) Insert 4 *A. M.*, between 193 and 443.

*Ans.*, 193, 243, 293, 343, 393, 443.

(**4.**) $\frac{1}{2} - \frac{2}{3} - \frac{11}{6} -$, etc. *Ans.*, $\frac{1}{6}(10 - 7n)$; and $\frac{n}{12}(13 - 7n)$.

(**5.**) $\frac{1}{4} + \frac{3}{8} + \frac{1}{2} +$, etc., to $n$ terms. *Ans.*, $s = \frac{n}{16}(n + 3)$.

(**6.**) $\frac{n-1}{n} + \frac{n-2}{n} + \frac{n-3}{n} +$, etc., to $n$ terms.

*Ans.*, $s = \frac{1}{2}(n - 1)$.

(**7.**) $\left(\frac{1}{a} - \frac{n}{x}\right) + \left(\frac{1}{a} - \frac{n-1}{x}\right) + \left(\frac{1}{a} - \frac{n-2}{x}\right)$, etc.

*Ans.*, $s = \frac{n}{a} - \frac{n}{x} \cdot \frac{n+1}{2}$.

(**8.**) How many strokes does a clock strike in 12 hours? *Ans.*, 78.

It will afford a good exercise for the student to solve the following cases on review, after having gone through Quadratics; though no importance need be attached to remembering the results, as the fundamental formulas

(1) $l = a + (n - 1)d$, and

(2) $s = \left[\frac{a+l}{2}\right]n$, are sufficient to resolve all cases.

| No. | Given. | Required. | Formulas. |
|---|---|---|---|
| 1. | $a$, $d$, $n$ | | $l = a + (n - 1)d$, |
| 2. | $a$, $d$, S | | $l = -\frac{1}{2}d \pm \sqrt{\{2dS + (a - \frac{1}{2}d)^2\}}$, |
| 3. | $a$, $n$, S | $l$ | $l = \frac{2S}{n} - a$, |
| 4. | $d$, $n$, S | | $l = \frac{S}{n} + \frac{(n-1)d}{2}$. |
| 5. | $a$, $d$, $n$ | | $S = \frac{1}{2}n\{2a + (n - 1)d\}$, |
| 6. | $a$, $d$, $l$ | S | $S = \frac{l + a}{2} + \frac{l^2 - a^2}{2d}$, |
| 7. | $a$, $n$, $l$ | | $S = (l + a)\frac{n}{2}$, |
| 8. | $d$, $n$, $l$ | | $S = \frac{1}{2}n\{2l - (n - 1)d\}$. |
| 9. | $a$, $n$, $l$ | | $d = \frac{l - a}{n - 1}$, |
| 10. | $a$, $n$, S | | $d = \frac{2(S - an)}{n(n - 1)}$, |
| 11. | $a$, $l$, S | $d$ | $d = \frac{l^2 - a^2}{2S - l - a}$, |
| 12. | $n$, $l$, S | | $d = \frac{2(nl - S)}{n(n - 1)}$. |
| 13. | $a$, $d$, $l$ | | $n = \frac{l - a}{d} + 1$, |
| 14. | $a$, $d$, S | | $n = \frac{\pm\sqrt{(2a - d)^2 + 8dS} - 2a + d}{2d}$, |
| 15. | $a$, $l$, S | $n$ | $n = \frac{2S}{l + a}$, |
| 16. | $d$, $l$, S | | $n = \frac{2l + d \pm \sqrt{(2l + d)^2 - 8dS}}{2d}$. |
| 17. | $d$, $n$, $l$ | | $a = l - (n - 1)d$, |
| 18. | $d$, $n$, S | | $a = \frac{S}{n} - \frac{(n - 1)d}{2}$, |
| 19. | $d$, $l$, S | $a$ | $a = \frac{1}{2}d \pm \sqrt{(l + \frac{1}{2}d)^2 - 2dS}$, |
| 20. | $n$, $l$, S | | $a = \frac{2S}{n} - l$. |

## GEOMETRICAL PROGRESSION.

(PAGES 299—303, *77—82.*)

See Complete Arithmetic (452—454).

**(1.)** Find the 9th term of the series, 1, 2, 4, 8, 16, etc. *Result*, 256.

**(2.)** Find the 8th term of the series, 2, 6, 18, etc. *Result*, 4374.

**(3.)** Find the 9th term of the series, 6, 18, 54, etc. *Result*, 39366.

**(4.)** Find the 6th term of the series, 2187, 729, etc. *Result*, 9.

**(5.)** Find the 8th term of the series, 81, 27, etc. *Result*, $\frac{1}{27}$.

**(6.)** Find the 9th term of the series, 256, 64, etc. *Result*, $\frac{1}{256}$.

**(7.)** Sum 8 terms of the series, 2, 6, 18, etc. *Sum*, 6560.

**(8.)** Sum 5 terms of the series, 2 4, 8, etc. *Sum*, 62.

**(9.)** Sum 8 terms of the series, 1, 4, 16, etc. *Sum*, 21845.

**(10.)** Sum 9 terms of the series, 1, 3, 9, etc. *Sum*, 9841.

**(11.)** Sum 6 terms of the series, 512, 128, 32, etc. *Sum*, 682½.

**(12.)** Sum 7 terms of the series, 15625, 3125, etc. *Sum*, 19531.

**(13.)** Sum 8 terms of the series, 147456, 36864, etc. *Sum*, 196605.

**(14.)** Sum 9 terms of the series, 19683, 6561. *Sum*, 29523.

**(15.)** Find the value of .5555 to infinity. *Value*, $\frac{5}{9}$.

**(16.)** Find the value of .6666, etc., to infinity. *Value*, $\frac{2}{3}$.

**(17.)** Find the value of .7777, etc., to infinity. *Value*, $\frac{7}{9}$.

**(18.)** Find the value of .212121, etc., to infinity. *Value*, $\frac{7}{33}$.

**(19.)** Find the value of .123123, to infinity. *Value*, $\frac{123}{999}$.

**(20.)** Find the value of .71333, to infinity. *Value*, $\frac{214}{300}$.

**(21.)** Find the sum of $1 + \frac{1}{x} + \frac{1}{x^2} + \frac{1}{x^3}$, etc. *Sum*, $\frac{x}{x-1}$.

**(22.)** Find the sum of $1 + \frac{1}{x+1} + \frac{1}{(x+1)^2}$. *Sum*, $\frac{x+1}{x}$.

**(23.)** Insert three geometrical means between the numbers 39 and 3159. *Series*, 39, 117, 351, 1053, 3159.

**(24.)** Insert four G. M. between 1 and 32. *Series*, 1, 2, 4, 8, 16, 32.

**(25.)** Insert three G. M. between 9 and $\frac{1}{9}$. *Series*, 9, 3, 1, $\frac{1}{3}$, $\frac{1}{9}$.

**(26.)** Insert four G. M. between 18 and 4374.
*Series*, 18, 54, 162, 486, 1458, 4374.

**(27.)** Insert three G. M. between 37 and 2997.

*Series*, 37, 111, 333, 999, 2997.

In a review, *after the pupil has been through the book*, it will be a good exercise for him to deduce the following formulas from the two fundamental ones.

| No. | Given. | Required. | Formulas. |
|---|---|---|---|
| 1. | $a,\ r,\ n$ | | $l = ar^{n-1}$, |
| 2. | $a,\ r,\ S$ | | $l = \dfrac{a + r(r-1)S}{r}$, |
| 3. | $a,\ n,\ S$ | $l$ | $l(S-l)^{n-1} - a(S-a)^{n-1} = 0$, |
| 4. | $r,\ n,\ S$ | | $l = \dfrac{(r-1)Sr^{n-1}}{r^n - 1}$. |
| 5. | $a,\ r,\ n$ | | $S = \dfrac{a(r^n - 1)}{r - 1}$, |
| 6. | $a,\ r,\ l$ | | $S = \dfrac{rl - a}{r - 1}$, |
| 7. | $a,\ n,\ l$ | $S$ | $S = \dfrac{\sqrt[n-1]{l^n} - \sqrt[n-1]{a^n}}{\sqrt[n-1]{l} - \sqrt[n-1]{a}}$, |
| 8. | $r,\ n,\ l$ | | $S = \dfrac{lr^n - l}{r^n - r^{n-1}}$. |
| 9. | $r,\ n,\ l$ | | $a = \dfrac{l}{r^{n-1}}$, |
| 10. | $r,\ n,\ S$ | $a$ | $a = \dfrac{(r-1)S}{r^n - 1}$, |
| 11. | $r,\ l,\ S$ | | $a = rl - (r-1)S$, |
| 12. | $n,\ l,\ S$ | | $a(S-a)^{n-1} - l(S-l)^{n-1} = 0$. |
| 13. | $a,\ n,\ l$ | | $r = \sqrt[n-1]{\dfrac{l}{a}}$, |
| 14. | $a,\ n,\ S$ | $r$ | $r^n - \dfrac{S}{a}r + \dfrac{S-a}{a} = 0$. |
| 15. | $a,\ l,\ S$ | | $r = \dfrac{S-a}{S-l}$, |
| 16. | $n,\ l,\ S$ | | $r^n - \dfrac{S}{S-l}r^{n-1} + \dfrac{l}{S-l} = 0$. |
| 17. | $a,\ r,\ l$ | | $n = \dfrac{log.l - log.a}{log.r} + 1$, |
| 18. | $a,\ r,\ S$ | $n$ | $n = \dfrac{log.[a + (r-1)S] - log.a}{log.r}$, |
| 19. | $a,\ l,\ S$ | | $n = \dfrac{log.l - log.a}{log.(S-a) - log.(S-l)} + 1$. |
| 20. | $r,\ l,\ S$ | | $n = \dfrac{log.l - log.[lr - (r-1)S]}{log.r} + 1$. |

## EXAMPLES IN APPLICATION OF PROPORTION AND PROGRESSION.

(PAGES 305—313.)

(**1.**) Divide the number 49 into two such parts, that the greater increased by 6, may be to the less diminished by 11, as 9 to 2.

*Parts*, 30, and 19.

(**2.**) What number is that, to which if 1, 5, and 13, be severally added, the first sum shall be to the second, as the second to the third?

*Ans.*, 3.

(**3.**) Find two numbers, the greater of which shall be to the less, as their sum to 42; and as their difference to 6.

SUG.—From $x : y :: x + y : 42$, and $x : y :: x - y : 6$, $x + y : x - y :: 42 : 6$, $2x : 2y :: 4 : 3$. $\therefore x = \frac{4}{3}y$. *Numbers*, 32, 24.

(**4.**) In a mixture of rum and brandy, the difference between the quantities of each, is to the quantity of brandy, as 100 is to the number of gallons of rum; and the same difference is to the quantity of rum, as 4 to the number of gallons of brandy. How many gallons are there of each? *Ans.*, 25 of rum, and 5 of brandy.

SUG. $x - y : y :: 100 : x$, and $x - y : x :: 4 : y$, give by dividing the former by the latter $1 : \frac{y}{x} :: 25 : \frac{x}{y}$, whence $1 : y^2 :: 25 : x^2$, or $1 : y :: 5 : x$, $5 : 5y :: 5 : x$. $\therefore x = 5y$; substituting in the second $4y : 5y :: 4 : y$. $\therefore y = 5$.

(**5.**) There are two numbers which are to each other as 3 to 2. If 6 be added to the greater and subtracted from the less, the sum and remainder will be to each other, as 3 to 1. What are the numbers?

*Ans.*, 24 and 16.

(**6.**) A bankrupt owed to two creditors 1400 dollars; the difference of the debts was to the greater as 4 to 9. What were the debts?

*Ans.*, 900, and 500 dollars.

(**7.**) Four places are situated in the order of the four letters A, B, C, D. The distance from A to D is 34 miles, the distance from A to B : distance from C to D :: 2 : 3, and one-fourth of the distance from A to B added to half the distance from C to D, is three times the distance from B to C. What are the respective distances?

*Ans.*, AB = 12, BC = 4, and CD = 18 miles.

(**8.**) A person having laid out a rectangular bowling-green, observed that if each side had been 4 yards longer, the adjacent sides would have been in the ratio of 5 to 4; but if each had been 4 yards shorter, the ratio would have been 4 to 3. What are the lengths of the sides?

*Ans.*, 36, and 28 yards.

(**9.**) A testator bequeathed his estate, amounting to $7830, to his three children, in such a manner that the share of the first was to that of the second as $2\frac{1}{2}$ to 2, and the share of the second to that of the third as $3\frac{1}{2}$ to 3. What were the shares?

*Ans.*, $3150, $2520, $2160.

(**10.**) To determine three numbers such that if 6 be added to the first and second, the sums will be in the ratio of 2 to 3; if 5 be added to the first and third, the sums will be in the ratio of 7 : 11; but if 36 be subtracted from the second and third, the remainders will be as 6 to 7. *Ans.*, 30, 48, 50.

Sug.—Let $2x - 6$, $3x - 6$, and $y$ be the numbers.

(**11.**) A, B, C, and D together have $3000; A's part is to B's as 2 to 3, B and C together have $1500, and C's part is to D's as 3 to 4. What is the sum possessed by each person?

*Ans.*, $500, $750, $750, $1000.

(**12.**) Each of two vessels contains a mixture of wine and water. A mixture consisting of equal measures from the two vessels, contains as much wine as water; and another mixture consisting of four measures from the first vessel and one from the second, is composed of wine and water in the ratio of 2 : 3. What are the relative amounts of wine and water in the vessels?

Sug.—Let $x$ and $y$ be proper fractions representing the part of each mixture respectively which is wine. Then $1 - x$ and $1 - y$ represent the fractional parts of each which are water. Now $a$ gallons from the first vessel contain $ax$ gallons of wine, and $a(1 - x)$ gallons of water; and $a$ gallons from the second contain $ay$ gallons of wine, and $a(1 - y)$ of water. $\therefore a(x + y) = a(1 - x + 1 - y)$, or $x + y = 1$, (1). In a similar manner we find that $4x + y : 4 - 4x + 1 - y :: 2 : 3$, or $4x + y = 2$, (2). Combining (1) and (2) we have $2x = y$; *i. e.*, there is twice as much wine in a gallon from the second vessel as in one from the first.

(**13.**) In a mass of zinc and copper, weighing 100 pounds, 8 parts are copper and 3 zinc. How much copper must be added in order that the ratio may be 14 : 5? *Ans.*, $\frac{2}{5}$ of a part, or $3\frac{7}{11}$ lbs.

Sug. $8+x:3::14:5$. $\therefore$ $x=\frac{2}{5}$. But $\frac{1}{11}$ is the part now copper. $\therefore$ $\frac{2}{5}$ of $\frac{1}{11}=\frac{2}{55}$ is the part to be added. $\frac{2}{55}$ of $100=3\frac{7}{11}$.

(**14.**) The range of a thermometer during the year was $44\frac{3}{10}°$. The ratio of the degrees at which it stood at the extreme points above and below zero was 7 : 4. What were the extreme points?

*Ans.*, $28\frac{21}{110}°$ above, and $16\frac{6}{55}°$ below.

(**15.**) Find two numbers having the ratio of 5 to 7, to which two other required numbers having the ratio of 3 to 5 being respectively added, the sums shall have the ratio of 9 to 13; and the difference of those sums shall be 16. *Numbers*, 30, 42, and 6, 10.

(**16.**) A merchant having mixed a certain number of gallons of brandy and water, found that if he had mixed 6 gallons more of each, he would have put into the mixture 7 gallons of brandy for every 6 of water; but, if he had mixed 6 less of each, he would have put in 6 gallons of brandy for every 5 of water. How many gallons of each did he mix? *Ans.*, 78 of brandy, and 66 of water.

(**17.**) From the first of two mortars in a battery 36 shells are thrown before the second is ready for firing. Shells are then thrown from both at the rate of 8 from the first to 7 from the second, the second mortar requiring as much powder for 3 charges as the first does for 4. After how many discharges of the second are the amounts of powder used by each the same? *Ans.*, 189.

(**18.**) A stage coach has 4 more outside than inside passengers. Seven outside can travel at $2 less expense for the journey than 4 inside. The fare of the whole amounts to $180; but at the end of half the journey the coach takes up 3 more outside and one more inside, in consequence of which the whole fare is increased in the ratio of 19 to 15. What is the number of passengers, and what is the fare of each kind?

*Ans.*, Passengers, 5 inside, 9 outside; fares, $18 and $10.

Sug. $2x$ being the fare for the whole distance, on the outside, and $2y$ on the inside, we have $14x+2=8y$. Again the whole fare for the last half the way would have been $90, but it was increased to $\$90+3x+y$. $\therefore$ $90+3x+y:90::19:15$, or $3x+y:6::4:1$. $\therefore$ $3x+y=24$, and $x=5$, $y=9$.

(**19.**) Divide the number $a$ into three such parts that the first may be to the second as $m$ to $n$, and the second part to the third as $p$ to $q$.

$$\textit{Parts},\ \frac{mpa}{mp+np+nq},\ \frac{npa}{mp+np+nq},\ \frac{nqa}{mp+np+nq}.$$

## PROBLEMS OF PURSUIT.

**(20.)** There are two places, 154 miles distant from each other, from which two persons, A and B, set out at the same instant, to meet on the road. A travels at the rate of 3 miles in 2 hours, and B at the rate of 5 miles in 4 hours. How long, and how far, did each travel before they met?

*Ans.*, 56 hours, and A travelled 84, and B, 70 miles.

**(21.)** A sets out express from C towards D, and three hours afterwards B set out from D towards C, travelling 2 miles an hour more than A. When they meet it appears that the distances they have travelled are in the proportion of 13 to 15 ; but had A travelled five hours less, and B had gone 2 miles an hour more, they would have been in the proportion of 4 : 5. How many miles did each go per hour, and how many hours did they travel before they met?

*Ans.*, A went 4, and B 6 miles an hour ; and they travelled 10 hours after B set out.

**(22.)** Two persons, A and B, set out from one place, and both go the same road, but A goes $a$ hours before B, and travels $n$ miles an hour ; B follows, and travels $m$ miles an hour. In how many hours, and in how many miles travel, will B overtake A?

*Ans.*, B travelled $\frac{na}{m-n}$ hours ; and A, $\frac{ma}{m-n}$ ; distance $\frac{mna}{m-n}$.

**(23.)** A person possesses a wagon with a mechanical contrivance by which the difference in the number of revolutions made by the fore and hind wheels may be determined. The fore wheel is $a$ feet, and the hind wheel is $b$ feet in circumference. What is the distance gone over, when the fore wheel has made $n$ revolutions more than the hind wheel?

*Ans.*, $\frac{abn}{b-a}$ feet.

**(24.)** The hour and minute hands of a watch are opposite at 6 o'clock ; when are they next opposite? *Ans.*, $5\frac{5}{11}$ m. past 7.

**(25.)** A person being asked the hour, answered that it was between five and six ; and the hour and minute hands were together. What was the time? *Ans.*, $27\frac{3}{11}$ m. past 5.

**(26.)** A and B start at the same time and travel in the same direction, A being 3 miles in advance. But A goes a mile in 20 minutes. and B, in 15. How long will it take B to overtake A?

SUG'S.—While A walks a mile (20 m.), B has gained 5 minutes' walk upon him. Now B has 45 minutes' walk to gain. $\therefore$ 20 — 15 : 45 :: 20 : $x$, and $x$ (the time) = 180.

Or, otherwise $\frac{x}{15} - \frac{x}{20} = 3$. $\therefore$ $x = 180$. This is the most simple process for this example; but the former will best illustrate the following examples. They may also be treated like examples 27—32, pp. 311—313 in the text-book.

(**27.**) The earth makes a revolution around the sun in about 365 days and Mars in about 687 days. How long is it from the time at which these planets are together on the same side of the sun till they are next together? That is, how long does it take the earth to gain an entire revolution?

*Ans.*, 687 — 365 : 365 :: 687 : $x$ $\therefore$ $x$ = 780 nearly.

[NOTE.—The time required by a planet to go around the sun is called its *Periodic Time*, and their being on the same side of the sun, at the same time is called *Conjunction*. The time from one conjunction to the next is the *Synodic Period*. The way in which this problem usually presents itself is this: We can *observe* when a planet is in conjunction with the earth, and thus knowing the time of the earth's revolution (a year), we can find the *Periodic Time* of the planet, or how long it takes to go around the sun.]

(**28.**) Calling the earth's *periodic time* 365¼ days, and observing Jupiter's *synodic period* to be about 398 days, how long is Jupiter in completing a revolution around the sun; that is, what is its *periodic time?* *Ans.*, $12\frac{2}{33}$ years.

(**29.**) Saturn's *synodic period* is about 378 days, what is its *periodic time?* *Ans.*, $29\frac{1}{18}$ years.

---

## PERCENTAGE AND INTEREST.

### (PAGES 314—326, *83—90.*)

[Examples upon the Business Rules of Arithmetic may be found in the Intellectual or Complete Arithmetic under the appropriate heads, and a few only will be inserted here.]

### (PAGE 314, *85.*)

See Intellectual Arithmetic, pages 143—146.

(Pages 317—322, ***88—90.***)

See Intellectual Arithmetic, pages 149—168, especially 164 to 168.

## PARTNERSHIP AND ALLIGATION.

(Pages 327—334, ***91—93.***)

See Complete Arithmetic, pages 241—250.

## MISCELLANEOUS.

(**1.**) Sold two lots for $1200 each; on one I gained 25%, on the other I lost 25%. Did I gain or lose by the operation, and how much?

*Ans.*, Lost $160.

(**2.**) Bought an article which I afterwards sold, gaining a certain per cent.; but if it had cost me 15% less, selling it for the same, I should have gained 30% more. What rate per cent. did I gain?

*Ans.*, 70%.

(**3.**) A commission merchant sold 2000 bbls. of flour at $6 per barrel, and invested the proceeds in sugar, after deducting his commission at 5% for selling, 2% for buying, and $47.40, the cost of shipping. How many barrels of sugar did he buy at $15 per barrel?

*Ans.*, 742.

(**4.**) A person owes a debt of $1500, towards the payment of which he is able to raise but $900. He offers this sum, to be applied in part to the payment of the debt, and in part to paying the interest, at 8 per cent., in advance, on his note at 12 months for the remainder. His creditor accepting, for what sum must the note be drawn?

*Ans.*, $652.17+.

(**5.**) A, B, and C formed a partnership; A advancing $5000 of the joint capital, B $4000, and C $3500. At the end of 6 months A withdrew $1500 from the business, B withdrew $500, and C increased his stock by ½ of its original amount. At the end of 12 months a dissolution occurred, when their profits amounted to $3765.12½: how should this sum have been divided between them?

*Ans.*, A should have had $1293.07+; B, $1140.94+; and C, $1331.10+.

(**6.**) A merchant took a farmer's note for $325.50, due, without interest, on the 1st of June, 1848, and some time afterwards the farmer got possession of a note against the merchant for \$500, due, without interest, on the 20th of January, 1850. Settlement was had on the 15th of August, 1849 ; how stood the matter of debt between them, allowing money to be at 7 per cent. ?

*Ans.*, \$132.51+ in favor of the farmer.

(**7.**) What is the present worth of a note of \$300, due 2 years 6 months hence, and bearing interest at 10%, when money is worth but 6%? *Ans.*, \$325.21+.

(**8.**) What is the worth of a note of \$100 dated, April 1st, 1861, and bearing interest at 12%, payable October 16th, 1866, on the 1st of July, 1864, money being worth 8%? *Ans.*, \$120.42+.

---

# *QUADRATIC EQUATIONS.*

## PURE QUADRATICS.

(Page 336, ***101.***)

(**1.**) Given $2x^2 + 130 = 5x^2 - 17$. $x = \pm 7$.

(**2.**) Given $51x^2 - 96 = 39x^2 + 96$. $x = \pm 4$.

(**3.**) Given $x^2 - \frac{25}{36}x^2 = 44$. $x = \pm 12$.

(**4.**) Given $\frac{3}{14x^2} + \frac{1}{2} = \frac{346}{686}$. $x = \pm 7$.

(**5.**) Given $\frac{1}{3}(7x^2+5) - \frac{1}{5}(16+4x^2)+6 = \frac{1}{2}(3x^2+9)$. $x = \pm 1$.

(**6.**) Given $\frac{1}{3}x^2 - 3 + \frac{5}{12}x^2 = \frac{7}{24} - x^2 + \frac{299}{24}$. $x = \pm 3$.

(**7.**) Given $5x^2 - 27 = 3x^2 + 215$. $x = \pm 11$.

(**8.**) Given $5x^2 - 1 = 244$. $x = \pm 7$.

(**9.**) Given $9x^2 + 9 = 3x^2 + 63$. $x = \pm 3$.

(**10.**) Given $2ax^2 + b - 4 = cx^2 - 5 + d - ax^2$.

$$x = \pm \sqrt{\frac{d - b - 1}{3a - c}}.$$

(**11.**) Given $x^2 + ab = 5x^2$. $x = \pm \frac{1}{2}\sqrt{ab}$.

(**12.**) Given $36x^2 = a^2$. $x = \pm \frac{1}{6}a$.

(**13.**) Given $\sqrt{4x^2+76}+100=120.$ $\qquad x=\pm 9.$

(**14.**) Given $4+\sqrt[3]{6x^2+3}=7.$ $\qquad x=\pm 2.$

(**15.**) Given $\sqrt[3]{10x^2+35}-1=4.$ $\qquad x=\pm 3.$

(**16.**) Given $x+\sqrt{x^2+a}=\dfrac{2a}{\sqrt{x^2+a}}.$ $\qquad x=\pm \frac{1}{3}\sqrt{3a}.$

(**17.**) Given $\sqrt[5]{9x^2-4}+6=8.$ $\qquad x=\pm 2.$

(**18.**) Given $\dfrac{\sqrt{x}+2a}{\sqrt{x}+b}=\dfrac{\sqrt{x}+4a}{\sqrt{x}+3b}.$ $\qquad x=\left(\dfrac{ab}{a-b}\right)^2.$

(**19.**) Given $\dfrac{ax-b^2}{\sqrt{ax}+b}=c+\dfrac{\sqrt{ax}-b}{c}.$ $\qquad x=\dfrac{1}{a}\cdot\left(b+\dfrac{c^2}{c-1}\right)^2.$

(**20.**) Given $4x^{\frac{2}{3}}=100.$ $\qquad x=\pm 125.$

(**21.**) Given $x^{\frac{2}{3}}=16.$ $\qquad x=64.$

(**22.**) Given $4x^{\frac{3}{4}}-50=6-3x^{\frac{3}{4}}.$ $\qquad x=16.$

(**23.**) Given $x^{\frac{2}{5}}=a.$ $\qquad x=a^{\frac{5}{2}}.$

(**24.**) Given $13x^{\frac{3}{2}}=338.$ $\qquad x=\sqrt[3]{(26)^2}.$

(**25.**) Given $\dfrac{a}{b+x}+\dfrac{a}{b-x}=c.$ $\qquad x=\pm\dfrac{1}{c}\sqrt{b^2c^2-2abc}.$

(**26.**) Given $x\sqrt{6+x^2}=1+x^2.$ $\qquad x=\pm \frac{1}{2}.$

(**27.**) Given $\dfrac{7x^2+8}{21}-\dfrac{x^2+4}{8x^2-11}=\dfrac{x^2}{3}.$ $\qquad x=\pm 2.$

(**28.**) Given $\sqrt{2+x}+\sqrt{x}=\dfrac{4}{\sqrt{2+x}}.$ $\qquad x=\frac{2}{3}.$

(**29.**) Given $\sqrt{5+x}+\sqrt{x}=\dfrac{15}{\sqrt{5+x}}.$ $\qquad x=4.$

(**30.**) $x+\sqrt{a^2+x^2}=\dfrac{2a^2}{\sqrt{a^2+x^2}}.$ $\qquad x=\pm\dfrac{a}{3}\sqrt{3}.$

(**31.**) Given $\dfrac{a}{x}+\dfrac{\sqrt{a^2-x^2}}{x}=\dfrac{x}{b}.$ $\qquad \pm\sqrt{2ab-b^2}.$

(32.) Given $\frac{\sqrt{x}+28}{\sqrt{x}+4}=\frac{\sqrt{x}+38}{\sqrt{x}+6}$. $x=4$.

(33.) Given $\sqrt{x+\sqrt{x}}-\sqrt{x-\sqrt{x}}=\frac{3}{2}\sqrt{\frac{x}{x+\sqrt{x}}}$. $x=\frac{25}{16}$.

(34.) Given $\sqrt{\frac{x}{4}+3}-\sqrt{\frac{x}{4}-3}=\sqrt{\frac{2x}{3}}$ $x=\pm 9\sqrt{2}$.

(35.) Given $\frac{1}{x+\sqrt{2-x^2}}+\frac{1}{x-\sqrt{2-x^2}}=ax$.

$x=\pm\frac{1}{a}\sqrt{a(a+1)}$.

(36.) Given $\frac{1}{1-\sqrt{1-x^2}}-\frac{1}{1+\sqrt{1-x^2}}=\frac{\sqrt{3}}{x^2}$. $x=\pm\frac{1}{2}$.

(37.) Given $\frac{\sqrt{a^2-x^2}-\sqrt{b^2+x^2}}{\sqrt{a^2-x^2}+\sqrt{b^2+x^2}}=\frac{c}{d}$.

$x=\pm\sqrt{\frac{a^2(d-c)^2-b^2(d+c)^2}{2(d^2+c^2)}}$.

(38.) Given $\frac{\sqrt{a+x}+\sqrt{a-x}}{\sqrt{x}}=\sqrt{\frac{x}{b}}$. $x=\pm 2\sqrt{ab-b^2}$.

---

## APPLICATIONS.

(PAGES 340—344.)

(1.) What two numbers are those whose difference is to the greater as 1 : 6, and whose difference multiplied by the less is 180?

*Ans.*, 36, and 30.

(2.) There are two numbers whose ratio is 2 : 3; and the difference of whose squares is 125. What are the numbers? *Ans.*, 10 and 15.

(3.) What two numbers are to each other as 3 : 7, the sum of whose squares is 232? Verify. *Ans.*, 6, and 14.

(4.) Find three numbers which are to each other as 2, 5, and 7, the sum of whose cubes is 12852. *Numbers*, 6, 15, and 21.

(5.) Find three numbers which are to each other as $m$, $n$, and $p$;

and such that the sum of the squares of the first two diminished by the square of the third is $a$.

*Numbers*, $\pm\sqrt{\frac{am^2}{m^2+n^2-p^2}}$, $\pm\sqrt{\frac{an^2}{m^2+n^2-p^2}}$, and $\pm\sqrt{\frac{ap^2}{m^2+n^2-p^2}}$.

(**6.**) A and B carried 100 eggs between them to market, and each received the same sum. If A had carried as many as B, he would have received 36 cents for them ; and if B had only taken as many as A, he would have received 16 cents. How many had each?

*Ans.*, A 40, and B 60.

(**7.**) Divide the number $s$ into two such parts that if $m^2$ be divided by the second and this quotient multiplied by the first, the product is the same as if $n^2$ be divided by the first and the quotient multiplied by the second. *Parts*, $\frac{ns}{m+n}$, and $\frac{ms}{m+n}$.

(**8.**) There are two numbers, which are in the proportion of 3 to 2 ; the difference of whose fourth powers is to the difference of their squares as 52 to 1. Required the numbers. *Ans.*, 6, and 4.

(**9.**) It is required to divide the number $a$ into two such parts, that the squares of those parts may be in the proportion of $m$ to $n$.

*Parts*, $\frac{a\sqrt{m} - a\sqrt{n}}{\sqrt{m}}$, and $\frac{a\sqrt{n}}{\sqrt{m}}$.

(**10.**) Some gentlemen made an excursion, and every one took the same sum. Each gentleman had as many servants attending him as there were gentlemen ; and the number of dollars which each had was double the number of all the servants ; and the whole sum of money taken out was $3456. How many gentlemen were there? *Ans.*, 12.

(**11.**) A detachment of soldiers from a regiment, being ordered to march on a particular service, each company furnished four times as many men as there were companies in the regiment ; but those becoming insufficient, each company furnished three more men ; when their number was found to be increased in the ratio of 17 to 16. How many companies were there in the regiment? *Ans.*, 12.

(**12.**) How long is a body in falling 800 feet?

*Ans.*, A little over 7 seconds.

(**13.**) A merchant laid out a certain sum in speculation, and found at the end of a year that he had gained $a$ dollars. This he added to his stock, and at the end of the second year found that he had gained exactly as much per cent. as in the year preceding. Proceeding in the same manner, and each year adding to his stock the gain of the preceding year, he found that at the end of the fourth year his stock was equal to eighty-one sixteenths of his original stock. What was his original stock? *Ans.*, $2a$ dollars.

(**14.**) From two towns, C and D, two travellers, A and B, set out at the same time to meet each other. It appeared, when they met, that B had gone 35 miles more than $\frac{3}{5}$ of the distance that A had travelled. A's rate was such that he would have made the whole distance in 20 hours and 50 minutes; but B would have required 30 hours. What was the distance from C to D? *Ans.*, 275 miles.

(**15.**) Two workmen, A and B, were engaged to work for a certain number of days at different rates. At the end of the time, A, who had played 4 of those days, had 75 shillings to receive; but B, who had played 7 of those days, received only 48 shillings. Now, had B only played 4 days, and A played 7 days, each would have received the same sum. For how many days were they engaged; and how many did each work, and what had each per day?

*Ans.*, They were engaged to work 19 days. A worked 15, and B 7 days. B received 4 shillings per day, and A 5 shillings per day.

(**16.**) A and B are two towns, situated on the bank of a river which runs at the rate of 4 miles per hour. A waterman rows from A to B and back again, and finds that he is 39 minutes longer upon the water than he would have been had there been no stream. The next day he repeats his voyage with another waterman, with whose assistance he can row half as fast again; and they find they are only 8 minutes longer in performing their voyage than they would have been had there been no stream. Determine the rate at which the waterman would row by himself. *Ans.*, 6 miles per hour.

(**17.**) A gentleman bought two pieces of silk, which together measured 36 yards. Each of them cost as many shillings by the yard, as there were yards in the piece, and their whole prices were as 4 to 1. What were the lengths of the pieces? *Ans.*, 24, and 12 yards.

(**18.**) A charitable person distributed a certain sum among some poor men and women, the numbers of whom were in the proportion of 4 to 5. Each man received one-third of as many shillings as there were persons relieved; and each woman received twice as many shillings as there were women more than men. Now the men received all together 18*s.* more than the women. How many were there of each?

*Ans.*, 12 men and 15 women.

(**19.**) There is a certain number such that if the square root of the product of the sum and difference of this number and $a$, be both added to and subtracted from $a$, the difference of the reciprocals of the results is $a$ divided by the square of the number. What is the number? *Ans.*, $\pm \frac{1}{3}a\sqrt{3}$.

(**20.**) The 4th root of the sum of the 4th power of $b$ and the 4th power of a certain number, is equal to the square root of the sum of the square of the number and $a$ square. What is the number?

*Ans.*, $\pm \frac{1}{2a}\sqrt{2(b^4 - a^4)}$.

---

## AFFECTED QUADRATICS.

### (Page 345, *103.*)

(**1.**) Given $6x + \frac{35 - 3x}{x} = 44$. $x = 7$, and $\frac{5}{6}$.

(**2.**) Given $\frac{4}{3}x^2 - x = \frac{85}{3}$. $x = 5$, and $-4\frac{1}{4}$.

(**3.**) Given $2x - \frac{37}{3} = -\frac{19}{x}$. $x = \frac{19}{6}$, and 3.

(**4.**) Given $x^2 - 4x + 10 = 65 + 2x$. $x = 11$, and $-5$.

(**5.**) Given $3x^2 - 27 - 6x = 14x - 2x^2 + 33$. $x = 6$, and $-2$.

(**6.**) Given $x^2 - 17x = -60$. $x = 12$, and 5.

(**7.**) Given $\frac{x^2}{81} - \frac{4x}{18} + 5 = 0$.

$x = 9(1 + 2\sqrt{-1})$, and $9(1 - 2\sqrt{-1})$.

(**8.**) Given $9x - 4x^2 + \sqrt{4x^2 - 9x + 11} = 5$. $x = \frac{1}{8}(9 \pm \sqrt{-31}$.

(**9.**) Given $3x^2 - 12x - 3 = 12x + 2 - 2x^2$. $x = 5$, and $-\frac{1}{5}$.

(**10.**) Given $5x^2 - 50x + 100 = 2x - 35$. $x = 5\frac{2}{5}$, and 5.

(**11.**) Given $7x^2 - 20x = 32$. $x = 4$, and $-\frac{8}{7}$.

(**12.**) Given $20x^2 - 10x = 10x - 3 - 5x^2$. $x = \frac{3}{5}$, and $\frac{1}{5}$.

**13.**) Given $8x^2 - 7x + 34 = 0$. $x = \frac{1}{16}(7 \pm \sqrt{-1039})$.

(**14.**) Given $3x^2 + x = 11$. $x = \frac{1}{6}(-1 \pm \sqrt{133})$.

(**15.**) Given $\frac{8-x}{2}-\frac{2x-11}{x-3}=\frac{x-2}{6}$. $x=6$, and $\frac{1}{2}$.

(**16.**) Given $\frac{7+x}{7-x}+\frac{7-x}{7+x}=\frac{29}{10}$. $x=7$, and $-3$.

(**17.**) Given $\frac{3x+5}{3x-5}-\frac{3x-5}{3x+5}=\frac{135}{176}$. $x=9$, and $-\frac{25}{81}$.

(**18.**) Given $\frac{6}{x+1}+\frac{2}{x}=3$. $x=2$, and $-\frac{1}{3}$.

(**19.**) Given $6x+\frac{35-3x}{x}=44$. $x=7$, and $\frac{5}{6}$.

(**20.**) Given $\frac{16}{x}-\frac{100-9x}{4x^2}=3$. $x=4$, and $2\frac{1}{12}$.

(**21.**) Given $3x-\frac{3x-10}{9-2x}=2+\frac{6x^2-40}{2x-1}$. $x=4$, and $\frac{23}{2}$.

(**22.**) Given $\frac{3x-2}{2x-5}+\frac{2x-5}{3x-2}=\frac{10}{3}$. $x=\frac{13}{3}$, and $\frac{1}{7}$.

(**23.**) Given $\frac{x+4}{x-3}-\frac{2x-3}{x+4}=\frac{59}{8}$. $x=4$, and $2\frac{57}{67}$.

(**24.**) Given $\frac{10}{x}-\frac{14-2x}{x^2}=2\frac{4}{9}$. $x=3$, and $1\frac{10}{11}$.

(**25.**) Given $3x-\frac{x^2}{4}+3=13$. $x=6\pm\sqrt{-4}=6\pm2\sqrt{-1}$.

(**26.**) Given $\frac{8-x}{2}=\frac{x-2}{6}+\frac{2x-11}{x-3}$. $x=6$, and $\frac{1}{2}$.

(**27.**) Given $mqx^2-mnx+pqx=np$. $x=\frac{n}{q}$, and $-\frac{p}{m}$.

(**28.**) Given $\frac{x-3a}{b}=\frac{9(b-a)}{x}$. $x=3b$, and $3(a-b)$.

(**29.**) Given $\frac{x+a}{x}-\frac{3x}{x-a}=0$. $x=\pm a\sqrt{-\frac{1}{2}}=\pm\frac{1}{2}a\sqrt{-2}$.

(**30.**) Given $\frac{1}{a+b+x}=\frac{1}{a}+\frac{1}{b}+\frac{1}{x}$. $x=-a$, and $-b$.

(**31.**) Given $\frac{a}{x-a}+\frac{b}{x-b}=\frac{2c}{x-c}$. $x=0$, and $\frac{2ab-ac-bc}{a+b-2c}$.

(**32.**) Given $\sqrt{x^2 - 4x - 9} = 6$. $x = 9$, and $-5$.

(**33.**) Given $\sqrt{3x^2 + 7x - 2} = 2$. $x = \frac{2}{3}$, and $-3$.

(**34.**) Given $x + \sqrt{5x + 10} = 8$. $x = 18$, and $3$.

(**35.**) Given $x - 2\sqrt{x^2 + x + 5} - 14 = 0$. $x = 4$, and $\frac{-44}{3}$.

[NOTE.—In verifying the last two examples it will be necessary to remember that the square root of a quantity is $\pm$. Thus, in 34, 3 verifies if we call $\sqrt{5x + 10}$ $+$, and 18, if we call it $-$. In 35, neither value satisfies the equation except for $\sqrt{x^2 + x + 5}$ $-$ ; whence the *form* of the equation is $x + 2\sqrt{x^2 + x + 5} - 14 = 0$. Such results are frequent, and should not cause surprise, because the *square root of a quantity is as really* $-$, *as* $+$.]

(**36.**) Given $2\sqrt{x} + \sqrt{2x + 1} = \frac{21}{\sqrt{2x + 1}}$. $x = -25$, and $4$.

Verify the result $x = -25$. See text-book, page 360, *Ex.* 2. *Ans.* to *Query.*

(**37.**) Given $\frac{x - \sqrt{x + 1}}{x + \sqrt{x + 1}} = \frac{5}{11}$. $x = 8$, and $-\frac{8}{9}$.

(**38.**) Given $\frac{\sqrt{4x} + 2}{4 + \sqrt{x}} = \frac{4 - \sqrt{x}}{\sqrt{x}}$. $x = 7\frac{1}{9}$, and $4$.

(**39.**)
- Given $x^2 - 10x = 10x - 96$. $x = 12$, and $8$.
- Given $3x^2 + x - 50 = 2x^2 + 2x + 160$. $x = 15$, and $-14$.
- Given $x^2 - 3x + 2 = \frac{5}{4} + \frac{1}{4}x$. $x = +3$, and $+\frac{1}{4}$.
- Given $2x^2 + 8x = 90$. $x = 5$, and $-9$.
- Given $3x^2 - 3x + 9 = 8\frac{1}{3}$. $x = \frac{1}{3}$, and $\frac{2}{3}$.
- Given $x^2 - \frac{5}{6}x + \frac{1}{6} = 0$. $x = \frac{1}{2}$, and $\frac{1}{3}$.

(**40.**) Given $x^{-2} + x^{-1} = 6$. $x = \frac{1}{2}$, and $-\frac{1}{3}$.

(**41.**) $\frac{15x - 5}{1 + 5\sqrt{x}} + \frac{2}{\sqrt{x}} = 3\sqrt{x}$. $x = 4$, and $\frac{1}{9}$.

(**42.**) Given $(x - c)\sqrt{ab} = (a - b)\sqrt{cx}$. $x = \frac{ac}{b}$, and $\frac{bc}{a}$.

(**43.**) Given $abx^2 - 2x(a + b)\sqrt{ab} = (a - b)^2$.

$$x = \frac{a + b \pm \sqrt{2(a^2 + b^2)}}{\sqrt{ab}}.$$

**(44.)** Given $\dfrac{123+41\sqrt{x}}{5\sqrt{x}-x}=\dfrac{20\sqrt{x}+4x}{3-\sqrt{x}}-\dfrac{2x^2}{(5\sqrt{x}-x)(3-\sqrt{x})}$.

$x=20\frac{1}{2}$, and 3.

**(45.)** Given $\dfrac{a+x+\sqrt{2ax+x^2}}{a+x-\sqrt{2ax+x^2}}=b$. $\quad x=-a\pm\sqrt{a^2+\dfrac{a^2(b-1)^2}{4b}}$.

---

(Page 355, ***106.***)

**(46.)** Given $3x^2+2x-9=76$. $\quad x=5$, and $-5\frac{2}{3}$.

**(47.)** Given $3x^2-\frac{3}{4}=5\frac{1}{4}+2x$. $\quad x=\frac{1}{3}\pm\frac{1}{3}\sqrt{19}$.

**(48.)** Given $\dfrac{x^2}{3}+\dfrac{4x}{5}-34\frac{1}{5}=0$. $\quad x=9$, and $-11\frac{2}{5}$.

**(49.)** Given $x^2-\dfrac{x}{2}=\dfrac{2}{3}-\dfrac{x}{3}$. $\quad x=\frac{1}{12}(1\pm\sqrt{97})$.

**(50.)** Given $x^2+\dfrac{17x}{4}=-\dfrac{17x}{4}-4$. $\quad x=-\frac{1}{2}$, and $-8$.

**(51.)** Given $3x^2-5x=50$. $\quad x=5$, and $-3\frac{1}{3}$.

**(52.)** Given $x+4=13-\dfrac{7x-8}{x}$. $\quad x=4$, and $-2$.

**(53.)** Given $5x-\dfrac{3x-3}{x-3}=2x+\dfrac{3x-6}{2}$. $\quad x=4$, and $-1$.

**(54.)** Given $(x-2)^2=\dfrac{x+\sqrt{x^2-9}}{x-\sqrt{x^2-9}}$. $\quad x=5$, and 3.

**(55.)** Given $abx^2+\dfrac{3a^2x}{c}=\dfrac{6a^2+ab-2b^2}{c^2}-\dfrac{b^2x}{c}$.

$x=\dfrac{2a-b}{ac}$ and $\dfrac{3a+2b}{bc}$.

**(56.)** Given $\dfrac{a-x}{x}+\dfrac{x}{a-x}=\dfrac{b}{c}$. $\quad x=\dfrac{2ac}{2c+b\pm\sqrt{b^2-4c^2}}$.

## EQUATIONS OF OTHER DEGREES SOLVED AS QUADRATICS.

(PAGE 357, ***107.***)

(**1.**) Given $x^3 = 125$. $x = 5$, and $\frac{5}{2}(-1 \pm \sqrt{-3})$.

SOLUTION.—Extracting the cube root $x = 5$. This is the only root given in this paragraph in the text-book. To obtain the other roots, transposing all the terms to the first member, we have $x^3 - 125 = 0$, and $x - 5 = 0$. Dividing $x^2 + 5x + 25 = 0$. Whence $x^2 + 5x = -25$, and $x = -\frac{5}{2} \pm \frac{5}{2}\sqrt{-3} = \frac{5}{2}(-1 \pm \sqrt{-3})$.

(**2.**) Given $x^3 = 1$. $x = 1$, and $\frac{1}{2}(-1 \pm \sqrt{-3})$.

(**3.**) Given $x^3 + 1 = 0$. $x = -1$, and $\frac{1}{2}(1 \pm \sqrt{-3})$.

(**4.**) Given $2x^{\frac{1}{5}} - 10 = x^{\frac{1}{5}} - 8$. $x = 32$.

(**5.**) Given $\frac{2}{x^{\frac{2}{3}}} + 1 = x^{-\frac{2}{3}} + \frac{50}{49}$. $x = 343$.

(**6.**) Given $\frac{x^4}{a^2} + \frac{2}{ab}x^4 = c^2 - \frac{x^4}{b^2}$. $x = \pm\sqrt{\pm\frac{abc}{a+b}}$.

(**7.**) Given $(x^{\frac{1}{2}} + 3a^2)^{\frac{1}{2}} - (x^{\frac{1}{2}} - 3a^2)^{\frac{1}{2}} = \frac{2a^{\frac{1}{2}}x^{\frac{1}{4}}}{b^{\frac{1}{2}}}$. $x = \frac{9a^3b^2}{4(b-a)}$.

(**8.**) Given $(x + 2)^{\frac{1}{m}} = (x^2 + 10x + 1)^{\frac{1}{2m}}$. $x = \frac{1}{2}$.

(**9.**) Given $y^{\frac{3}{4}} = 64$, $y = 256$. $y^{\frac{2}{3}} = 49$, $y = \pm 343$.

Also $y^{\frac{5}{2}} = 32$, $y = 4$.

(PAGE 358, ***108.***)

(**1.**) Given $3x^{\frac{4}{3}} - \frac{5x^{\frac{8}{3}}}{2} + 592 = 0$. $x = 8$, and $(-\frac{74}{5})^{\frac{3}{4}}$.

(**2.**) Given $x^{\frac{6}{5}} + x^{\frac{3}{5}} = 756$. $x = 243$, and $(-28)^{\frac{5}{3}}$.

(**3.**) Given $x^3 - x^{\frac{3}{2}} = 56$. $x = 4$, and $(-7)^{\frac{2}{3}}$.

(**4.**) Given $x = \frac{\sqrt{x^4 - a^4}}{a}$. $x = \pm a\sqrt{\frac{1 \pm \sqrt{5}}{2}}$.

**(5.)** Given $x^{\frac{5}{2}} - x = 56x^{-\frac{1}{2}}$. $x = 4$, and $\sqrt[3]{49}$.

**(6.)** Given $\sqrt{x^5} + \sqrt{x^3} = 6\sqrt{x}$. $x = 2$, and $-3$.

**(7.)** Given $x\sqrt{x} - 2\sqrt{x} = x$. $x = 4$, and $+1$.

**(8.)** Given $x^2 + 11 + \sqrt{(x^2+11)} = 42$. $x = \pm 5$, and $\pm\sqrt{38}$.

**(9.)** Given $3x^2 + x^{\frac{7}{6}} - 3104\,x^{\frac{1}{3}} = 0$. $x = 64$, and $\left(-\frac{97}{3}\right)^{\frac{6}{5}}$.

**(10.)** Given $x - \sqrt{x} = 20$. $x = 25$, and $16$.

**(11.)** Given $ax^{\frac{3}{2}} + bx^{\frac{3}{4}} = c$ $x = \sqrt[3]{\frac{(-b \pm \sqrt{b^2+4ac})^4}{16a^4}}$.

**(12.)** Given $\frac{x+\sqrt{x}}{x-\sqrt{x}} = \frac{x^2-x}{4}$. $x = 4$, $1$, and $\frac{-3 \pm \sqrt{-7}}{2}$.

**(13.)** Given $x^n - 2ax^{\frac{1}{2}n} = b$. $x = (a \pm \sqrt{a^2+b})^{\frac{2}{n}}$.

## (PAGE 360, *109.*)

**(1.)** Given $(2x^2 - 4x + 1)^2 - (2x^2 - 4x + 1) = 42$.
$x = 3$, $-1$, and $1 \pm \sqrt{-\frac{5}{2}}$.

**(2.)** Given $(x+2)^2 = 20 - (x + 2)$. $x = -7$, and $2$.

**(3.)** Given $(x^2 - x)^2 - x^2 = 132 - x$. $x = 4$, $-3$, and $\frac{1}{2} \pm \frac{\sqrt{-43}}{2}$.

**(4.)** Given $\frac{8}{(2x-4)^2} = 1 + \frac{16}{(2x-4)^4}$. $x = 3$, and $1$.

**(5.)** Given $x + 3 = \sqrt{x+5} + 4$. $x = 4$, and $-1$.

**(6.)** Given $3x^2 + 1 + (3x^2+1)^{\frac{1}{2}} = 56$. $x = \pm 4$, and $\pm\sqrt{21}$.

**(7.)** Given $x^2 - 150 - 16x = 5\sqrt{x^2 - 16x}$. $x = 25$, and $-9$.

**(8.)** Given $x^2 - 7x + \sqrt{x^2 - 7x + 18} = 24$. $x = 9$, and $-2$.

**(9.)** Given $9x - 2x^2 - 2 = 3 + 2x^2 - \sqrt{4x^2 - 9x + 11}$.
$x = 2$, and $\frac{1}{4}$.

**(10.)** Given $x^4\left(1 + \frac{1}{3x}\right)^2 - (3x^2 + x = 70$.
$x = 3$, $-3\frac{1}{3}$, and $\frac{1}{6}(-1 \pm \sqrt{-251})$.

(PAGE 361, ***110.***)

**(11.)** Given $x^4 - 2x^3 + x = 30$. $x = 3, -2$, and $\frac{1}{2}(1 \pm \sqrt{-19})$.

**(12.)** Given $\frac{x^4}{10} + \frac{7}{2}x^2 + \frac{12}{5} = x^3 + 5x$. $x = 1, 2, 3$, and $4$.

**(13.)** Given $x^3 - 9x^2 + 23x = 15$. $x = 1, 3$, and $5$.

**(14.)** Given $x^3 - 8x^2 + 19x - 12 = 0$. $x = 1, 3$, and $4$.

**(15.)** Given $x^4 + 2ax^3 + 5a^2x^2 + 4a^3x = d$.

$$x = -\frac{a}{2} \pm \frac{1}{2}\sqrt{-7a^2 \pm 4\sqrt{4a^4 + d}}.$$

(PAGE 363, ***111.***)

**(1.)** Given $x^3 - 39x = -70$. $x = 2, 5$, and $-7$.

**(2.)** Given $x^3 - 21x = -344$. $x = -8$, and $4 \pm 3\sqrt{-3}$.

**(3.)** Given $\dfrac{x^2 + 3x - 7}{x + 2 + \dfrac{18}{x}} = 1$. $x = 3, -2$, and $-3$.

**(4.)** Given $x^4 + 2x^3 - 7x^2 - 8x = -12$. $x = 1, 2, -3$, and $-2$.

**(5.)** Given $x^4 - 12x^3 + 47x^2 - 72x + 36 = 0$. $x = 1, 2, 3$, and $6$.

**(6.)** Given $x^3 + 3x^2 - 6x = 8$. $x = 2, -1$, and $-4$.

**(7.)** Given $\frac{x^4}{2} + \frac{17x^3}{4} - 17x = 8$. $x = \pm 2, -\frac{1}{2}$, and $-8$.

**(8.)** Given $x^4 - 36x^2 + 72x - 36 = 0$.

$x = 3 \pm \sqrt{3}$, and $-3 \pm \sqrt{15}$.

**(9.)** Given $x^4 - 6x^2 - 8x - 3 = 0$. $x = 3, -1, -1$, and $-1$.

**(10.)** Given $x^4 - 9x^2 + 4x + 12 = 0$. $x = 2, 2, -1$, and $-3$.

**(11.)** Given $y^4 - 4y^2 = y^2 + 8y + 12$.

$y = 3, -2$, and $\frac{1}{2}(-1 \pm \sqrt{-7})$.

**(12.)** Given $x^4 - 8x^3 + 24x^2 - 32x + 16 = 81$.

$x = 5, -1$, and $2 \pm 3\sqrt{-1}$.

**(13.)** Given $\dfrac{x^2 + 8}{6} - \dfrac{4x + 6}{x^2} = \dfrac{8 - \dfrac{2}{x^2}}{3}$.

$x = 4, -2$, and $-1 \pm \sqrt{-3}$.

**(14.)** Given $\sqrt[6]{\frac{1}{x^4}} + \sqrt[3]{\frac{1}{x}} = \frac{3 - \sqrt[3]{x^2}}{x}$. $\qquad x = \pm 1$, and $\pm \frac{27}{8}$.

SUG.—This equation may be written $\frac{1}{x^{\frac{2}{3}}} + \frac{1}{x^{\frac{1}{3}}} = \frac{3 - x^{\frac{2}{3}}}{x}$. Whence $x^{\frac{1}{3}} + x^{\frac{2}{3}} = 3 - x^{\frac{2}{3}}$; or $2x^{\frac{2}{3}} + x^{\frac{1}{3}} = 3$.

---

## (PAGE 367, *112.*)

**(1.)** Given $x - y = 12$, and $x^2 + y^2 = 74$. $\qquad x = 7, 5$; $y = -5, -7$.

**(2.)** Given $x + y = 4$, and $\frac{1}{x} + \frac{1}{y} = 1$. $\qquad x = 2$, $y = 2$.

**(3.)** Given $x + 4y = 14$, and $4x - 2y + y^2 = 11$.
$x = 2, -46$; $y = 3, 15$.

**(4.)** Given $x + \frac{y}{2} = 11$, and $xy + 2y^2 = 120$.
$x = 8, 17\frac{2}{3}$; $y = 6, -13\frac{1}{3}$.

**(5.)** Given $x - \frac{x-y}{2} = 4$, and $y - \frac{x+3y}{x+2} = 1$. $\qquad x = 2, 5$; $y = 6, 3$.

**(6.)** Given $\frac{10x+y}{xy} = 3$, and $9y - 9x = 18$. $\qquad x = 2, -\frac{1}{3}$; $y = 4, \frac{5}{3}$.

**(7.)** Given $2x - 3y = 1$, and $2x^2 + xy - 5y^2 = 20$.
$x = 5, -9\frac{1}{4}$; $y = 3, -6\frac{1}{2}$.

**(8.)** Given $3x + y = 18$, and $x^2 + 2y^2 - 40 = 3$.
$x = 5, \frac{121}{19}$; $y = 3, -\frac{21}{19}$.

**(9.)** Given $4(x + y) = 5$, and $8xy = 3$. $\qquad x = \frac{1}{2}, \frac{3}{4}$; $y = \frac{3}{4}, \frac{1}{2}$.

---

## (PAGE 368, *114.*)

**(1.)** Given $x^2 + xy = 15$, and $xy - y^2 = 2$.
$x = \pm 3, \pm \frac{5}{2}\sqrt{2}$; $y = \pm 2, \pm \frac{1}{2}\sqrt{2}$.

**(2.)** Given $x^2 + xy = 12$, and $xy - 2y^2 = 1$.
$x = \pm 3, \pm \frac{4}{3}\sqrt{6}$; $y = \pm 1, \pm \frac{1}{6}\sqrt{6}$.

**(3.)** Given $2x^2 + 3xy + y^2 = 20$, and $5x^2 + 4y^2 = 41$.
$x = \pm \frac{13}{2}\sqrt{\frac{164}{861}}, \pm 1$; $y = \pm \sqrt{\frac{164}{861}}, \pm 3$.

(**4.**) Given $x^2 + 4y^2 = 256 - 4xy$, and $3y^2 - x^2 = 39$.
$x = \pm 6, \pm 102$ ; $y = \pm 5, \mp 59$.

(**5.**) Given $x^2 - xy = 6$, and $x^2 + y^2 = 61$.
$x = \pm 6, \mp \frac{1}{2}\sqrt{2}$ ; $y = \pm 5, \pm \frac{11}{2}\sqrt{2}$.

(**6.**) Given $3x^2 + xy = 68$, and $4y^2 + 3xy = 160$.
*Rational roots*, $x = \pm 4$ ; $y = \pm 5$.

(**7.**) Given $6x^2 + 2y^2 = 5xy + 12$, and $2xy + 3x^2 = 3y^2 - 3$.
*Rational roots*, $x = \pm 2$ ; $y = \pm 3$.

(**8.**) Given $4x^2 - 2xy = 12$, and $2y^2 + 3xy = 8$.
$x = \pm 2, \mp 3\sqrt{\frac{1}{7}}$ ; $y = \pm 1, \pm 8\sqrt{\frac{1}{7}}$.

---

## (Page 370, *115.*)

(**1.**) Given $x^2 + y^2 - x - y = 18$, and $xy + x + y = 19$.
$x = 4, 3$, and $-4 \pm \sqrt{-11}$ ; $y = 3, 4$, and $-4 \mp \sqrt{-11}$.

(**2.**) Given $x^3 + y^3 = 35$, and $x + y = 5$. $x = 3, 2$ ; $y = 2, 3$.

(**3.**) Given $\frac{x+y}{x-y} + \frac{x-y}{x+y} = \frac{5}{2}$, and $x^2 + y^2 = 20$.
$x = \pm 3\sqrt{2}$ ; $y = \pm \sqrt{2}$.

(**4.**) Given $xy(x + y) = 30$, and $x^3 + y^3 = 35$. $x = 3, 2$ ; $y = 2, 3$.

Sug.—Putting $x = m + n$, and $y = m - n$, and subtracting the first equation from the second, there results $8mn^2 = 5$. From this $2mn^2 = \frac{5}{4}$, which substituted in the first gives $2m^3 = \frac{125}{4}$. $\therefore$ $m = \frac{5}{2}$, and $n = \pm \frac{1}{2}$.

(**5.**) Given $x^2y + y^2x = 20$, and $\frac{1}{x} + \frac{1}{y} = \frac{5}{4}$. $x = 1, 4$ ; $y = 4, 1$.

Sug. $x = m + n$, $y = m - n$. The first becomes $(m + n)^2(m - n) + (m - n)^2(m + n) = 2$, and the second $(m + n)(m - n) = \frac{8}{5}m$. Substituting this value of $(m + n)(m - n)$ in the preceding, it becomes $(m + n)\frac{8}{5}m + (m - n)\frac{8}{5}m = 20$, or $16m^2 = 100$. $\therefore$ $m = \pm \frac{5}{2}$.

(**6.**) Given $x + y = 11$, and $x^4 + y^4 = 2657$. $x = 4, 7$ ; $y = 7, 4$.

(**7.**) Given $x + y = 10$, and $x^5 + y^5 = 17050$. $x = 3, 7$ ; $y = 7, 3$.

(**8.**) Given $x^2 + y^2 = 7 + xy$, and $x^3 + y^3 = 6xy - 1$.
$x = 2, 3$ ; $y = 3, 2$.

Sug.—By the same substitution as above, we have, from the first $3n^2 = 7 - m^2$; and, from the second, $2m^3 + 2m(3n^2) = 6m^2 - 6n^2 - 1$.

Substituting $7-m^2$ for $3n^2$ in the last, it becomes $2m^3+14m-2m^3=6m^2-14+2m^2-1$, or $8m^2-14m=15$.

(**9.**) Given $x+y=72$, and $\sqrt[3]{x}+\sqrt[3]{y}=6$. $x=8, 64$; $y=64, 8$.

SUG.—Put $\sqrt[3]{x}=m+n$, and $\sqrt[3]{y}=m-n$. $\therefore$ $m=3$.

(**10.**) Given $x^5-y^5=3093$, and $x-y=3$.
$x=5, -2$, $y=2, -5$.

SUG.—Divide the first by the second and then make the ordinary substitution in the resulting equation and the second.

(**11.**) Given $x+y=10$, and $\sqrt{\frac{x}{y}}+\sqrt{\frac{y}{x}}=\frac{5}{2}$.
$x=2, 8$; $y=8, 2$.

SUG.—The second equation cleared of fractions is $2(x+y)=5\sqrt{xy}$. Put $\sqrt{x}=m+n$, and $\sqrt{y}=m-n$.

A more elegant solution is to substitute the value of $x+y$, in the second after clearing it of fractions, thus obtaining $4=\sqrt{xy}$, or $xy=16$. Subtracting 4 times this from the square of the first, $x^2-2xy+y^2=36$. $\therefore$ $x-y \pm 6$.

---

## (PAGE 372, *116.*)

(**1.**) Given $x^{\frac{3}{2}}+y^{\frac{2}{3}}=3x$, and $x^{\frac{1}{2}}+y^{\frac{1}{3}}=x$. $x=4, 1$; $y=8$.

SUG.—Put $x^{\frac{1}{2}}=m$, and $y^{\frac{1}{3}}=n$; whence the equations become $m^3+n^2=3m^2$, and $m+n=m^2$. From the latter, $n^2=m^4-2m^3+m^2$, which substituted in the former gives $m^2-m=2$. But $m^2-m=n$. $\therefore$ $n=2$.

(**2.**) Given $x^{\frac{1}{2}}+y^{\frac{1}{2}}=4$, and $x^{\frac{3}{2}}+y^{\frac{3}{2}}=28$. $x=9, 1$; $y=1, 9$.

(**3.**) Given $x^{\frac{2}{3}}+y^{\frac{2}{3}}+2x^{\frac{1}{3}}+2y^{\frac{1}{3}}=23$, and $x^{\frac{2}{3}}y^{\frac{2}{3}}=36$.
$x=27, 8$; $y=8, 27$.

SUG.—By adding twice the second to the first, it becomes $(x^{\frac{1}{3}}+y^{\frac{1}{3}})^2+2(x^{\frac{1}{3}}+y^{\frac{1}{3}})=95$.

(**4.**) Given $x^{\frac{1}{3}}+y^{\frac{1}{3}}=8$, and $x^{\frac{2}{3}}+y^{\frac{2}{3}}+x^{\frac{2}{3}}y^{\frac{2}{3}}=259$.
$x=125, 27$; $y=27, 125$.

SUG.—The unknown quantities are similarly involved. See ***115.*** Also by subtracting the square of the first from the second $x^{\frac{2}{3}}y^{\frac{2}{3}}$ —

$2x^{\frac{1}{3}}y^{\frac{1}{3}} = 195$. $\therefore$ $x^{\frac{1}{3}}y^{\frac{1}{3}} = 15$, and $-13$. From the square of the first subtracting 4 times this $x^{\frac{2}{3}} - 2x^{\frac{1}{3}}y^{\frac{1}{3}} + y^{\frac{2}{3}} = 4$. $\therefore$ $x^{\frac{1}{3}} - y^{\frac{1}{3}} = \pm 2$.

(**5.**) Given $\frac{x}{y} + \frac{4x^{\frac{1}{2}}}{y^{\frac{1}{2}}} = \frac{33}{4}$ and $x - y = 5$. $x = 9$, $y = 4$.

SUG.—See *Ex.* 8 in text-book.

---

## MISCELLANEOUS.

[NOTE.—The following are arranged without reference to the corresponding examples in the text-book, but as further illustrations of *Special Expedients.*]

(**1.**) $x^2(x + y) = 80$, and $x^2(2x - 3y) = 80$. $x = 4$, $y = 1$.

SUG.—Put the first members equal and divide by $x^2$.

(**2.**) $\left(3 - \frac{6y}{x+y}\right)^2 + \left(3 + \frac{6y}{x-y}\right)^2 = 82$, and $xy = 2$.

$x = \pm 2, \pm 1$; $y = \pm 1, \pm 2$.

SUG.—From the first $9\left(\frac{x-y}{x+y}\right)^2 + 9\left(\frac{x+y}{x-y}\right)^2 = 82$. Which reduces as follows $(x-y)^4 + (x+y)^4 = \frac{82}{9}(x^2 - y^2)^2$, $9x^4 + 54x^2y^2 + 9y^4 = 41x^4 - 82x^2y^2 + 41y^4$, $32x^4 + 32y^4 = 136x^2y^2$. But $x^2y^2 = 4$. Whence $x^4 + y^4 = 17$.

(**3.**) $\sqrt{\frac{x}{y}} + \sqrt{\frac{y}{x}} = \frac{7}{\sqrt{xy}} + 1$, and $\sqrt{x^3y} + \sqrt{y^3x} = 78$.

SUG.—Clearing the first of fractions, $x + y = 7 + \sqrt{xy}$. Factoring the second $x\sqrt{xy} + y\sqrt{xy} = 78$; whence $x + y = \frac{78}{\sqrt{xy}}$, and $\sqrt{xy} = \frac{78}{x+y}$. Substituting the first of these in the former $\frac{78}{\sqrt{xy}} = 7 + \sqrt{xy}$, or $78 = 7\sqrt{xy} + xy$. This solved as an affected quadratic in terms of $\sqrt{xy}$ gives $\sqrt{xy} = 6$, and $-13$. In a similar manner $x + y = 7 + \frac{78}{x+y}$, or $(x+y)^2 - 7(x+y) = 78$. Whence $x + y = 13$, and $-6$. $\therefore$ $x = 4, 9$; $y = 9, 4$.

(**4.**) $x + y = 10$, and $\sqrt{\frac{x}{y}} + \sqrt{\frac{y}{x}} = \frac{5}{2}$. $x = 2, 8$ ; $y = 8, 2$.

(**5.**) $x^4 + y^4 = 14x^2y^2$, and $x + y = a$.

$$x = \frac{a}{2}(1 \pm \sqrt{3}), \frac{a}{2}(1 \pm \tfrac{1}{3}\sqrt{3}) ; \; y = \frac{a}{2}(1 \mp \sqrt{3}), \frac{a}{2}(1 \mp \tfrac{1}{3}\sqrt{3}).$$

SUG.—Add $2x^2y^2$ to both members of the first equation, and then extract the square root; whence $x^2 + y^2 = 4xy$.

(**6.**) $x + \sqrt{x^2 - y^2} = 8$, and $x - y = 1$. $x = 5, 13$ ; $y = 4, 12$.

SUG. $x + \sqrt{(x+y)(x-y)} = 8$. $x + \sqrt{x+y} = 8$. $x + y = (8 - x)^2$, or $2x + 1 = (8 - x)^2$.

(**7.**) $x + y + \sqrt{xy} = 14$, and $x^2 + y^2 + xy = 84$.
$x = 8, 2$ ; $y = 2, 8$.

(**8.**) $x + y - \sqrt{xy} = 7$, and $x^2 + y^2 + xy = 133$.
$x = 9, 4$ ; $y = 4, 9$.

SUG.—The 7th and 8th are symmetrical (***115***).

(**9.**) $x^2y - xy = 6$, and $x^3y - y = 21$. $x = 2, \frac{1}{2}$ ; $y = 3, -24$.

SUG.—Dividing the first by the second $\frac{x^2y - xy}{x^3y - y} = \frac{6}{21}, \frac{x^2 - x}{x^3 - 1} = \frac{6}{21}$, $\frac{x}{x^2 + x + 1} = \frac{6}{21}$.

(**10.**) $x^3 + xy^2 = 39$, and $x^2y + y^3 = 26$. $x = 3$ ; $y = 2$.

(**11.**) $xy = 8$, and $\frac{x^2}{y} + \frac{y^2}{x} = 9$. $x = 4, 2$ ; $y = 2, 4$.

SUG. $x^3 + y^3 = 72$, or $x^6 + x^3y^3 = 72x^3$. Substitute for $x^3y^3$ its value 512, and $x^6 - 72x^3 = -512$.

(**12.**) $xy = 25$, and $x^3 + y^3 = 10xy$. $x = 5$ ; $y = 5$.

SUG.—This is similar to the 11th.

## MISCELLANEOUS EXAMPLES IN APPLICATION OF THE PRINCIPLES OF QUADRATICS.

(**1.**) To divide the number 56 into two such parts, that their product shall be 640. *Parts*, 40, and 16.

(**2.**) There are two numbers whose difference is 7, and half their product *plus* 30, is equal to the square of the lesser number. What are the numbers? *Ans.*, 12, and 19.

(**3.**) A and B set out at the *same time* to a place at the distance of 150 miles. A travelled 3 miles an hour faster than B, and arrives at his journey's end 8 hours and 20 minutes before him. At what rate did each person travel per hour? *Ans.*, A 9, and B 6 miles per hour.

(**4.**) The difference of two numbers is 6; and if 47 be added to *twice the square of the lesser*, it will be equal to the *square of the greater*. What are the numbers? *Ans.*, 17, and 11.

(**5.**) There are two numbers whose product is 120. If 2 be added to the lesser, and 3 subtracted from the greater, the product of the sum and remainder will also be 120. What are the numbers? *Ans.*, 15, and 8.

(**6.**) Required to find two numbers such, that their product added to the square of the less is 10, and the product added to twice the square of the greater is 24. *Numbers*, 2, and 3.

(**7.**) Required to find two numbers such, that their sum multiplied by the greater is 77, and their difference multiplied by the less is equal to 12. *Numbers*, 7, and 4.

(**8.**) Find two such numbers, that the sum of their squares shall be 5, and three times their product shall be 6. *Numbers*, 1, and 2.

(**9.**) Required two numbers such, that four times the square of the greater, diminished by twice their product, is equal to 12, and twice the square of the less, added to three times their product, is equal to 8. *Numbers*, 1, and 2.

(**10.**) Find two numbers such, that their sum, quotient, and differences of their squares shall be equal to each other.

*Numbers*, $\frac{1}{2}(2+\sqrt{2})$, and $\frac{1}{2}\sqrt{2}$.

(**11.**) There is a number consisting of 2 digits, which, when divided by the sum of its digits, gives a quotient greater by 2 than the first digit. But if the digits be inverted, and then divided by a number greater by unity than the sum of the digits, the quotient is greater by 2 than the preceding quotient. Required the number.

*Number*, 24.

(**12.**) What two numbers are those, whose product is 24, and whose sum added to the sum of their squares is 62? *Ans.*, 4, and 6.

(**13.**) Divide 30 into two such parts, that their product shall be equal to 8 times their difference. *Parts*, 6, and 24.

(**14.**) A person bought a certain number of sheep for $120. If there had been 8 more for the same sum, each sheep would have cost him $0.50 less. How many sheep did he buy? *Ans.*, 40.

(**15.**) Find two numbers such, that their sum shall be 8, and the sum of their 4th powers 706. *Numbers*, 5, and 3.

(**16.**) The sum of two numbers is 7, and the sum of their 5th powers 3157. Required the numbers. *Numbers*, 5, and 2.

(**17.**) The difference between two numbers is 1, and the difference of their 5th powers is 211. Required the numbers.

*Numbers*, 2, and 3.

(**18.**) Find two numbers such, that their product, difference of their squares, and quotient of their cubes, shall all be equal to each other.

*Numbers*, $\frac{1}{2}(3 + \sqrt{5})$, and $\frac{1}{2}(1 + \sqrt{5})$.

(**19.**) Divide 21 and 35 into two such parts, that the difference of the squares of the first parts may be 57, and that of the second parts 407. *Parts*, 8, 13, and 11, 24.

(**20.**) Required two numbers such, that their sum, multiplied by the sum of their squares, is 272, and their difference, multiplied by the difference of their squares, is 32. *Numbers*, 5 and 3.

(**21.**) There are three numbers, the difference of whose differences is 3; their sum is 21; and the sum of the squares of the greatest and least is 137. Required the numbers. *Numbers*, 4, 6, and 11.

(**22.**) There are two numbers, such, that if the lesser be taken from three times the greater, the remainder will be 35; and if four times the greater be divided by three times the lesser *plus* one, the quotient will be equal to the lesser number. What are the numbers?

*Ans.*, 13, and 4.

(**23.**) Find the time between 3 and 4 o'clock, when the hour and minute hands are at right angles to each other.

*Time*, $32\frac{8}{11}$ min. past 3.

(**24.**) Two vessels each contain a mixture of wine and water; one in the ratio of 3 : 2, the other in the ratio of 7 : 3. What quantity must be taken from each in order to make a third mixture, which shall contain 5 gallons of water to 11 of wine?

*Result*, 2 gallons from one, and 14 from the other.

(**25.**) A man rows a distance of 21 miles, and back again, in 8 hours, and finds that he can row 7 miles with the stream for 5 against it. Required the rate of the stream, and the time of rowing each way.

*Result*, Rate of stream, $\frac{9}{10}$ mile per hour; $3\frac{1}{3}$ hours in going, $4\frac{2}{3}$ in returning.

(**26.**) A and B set out from two towns which were at the distance of 247 miles, and travelled the direct road till they met. A went 9 miles a day; and the number of days, at the end of which they met, was greater by 3 than the number of miles which B went in a day. How many miles did each go? *Ans.*, A 117, and B 130 miles.

(**27.**) A, B, and C possess certain sums of money, such that if A receive $\frac{1}{2}$ of B's and C's, he will possess $a$ dollars; if B received $\frac{1}{3}$ of A's and C's, he will possess $b$ dollars; and if C receives $\frac{1}{4}$ of A's and B's, he will possess $c$ dollars. Required the original sums of each.

*Sums*, $\frac{1}{17}(22a-9b-8c)$; $\frac{1}{17}(21b-4c-6a)$; $\frac{1}{17}(20c-3b-4a)$.

(**28.**) A certain quantity of wine was drawn from a full vessel, containing 81 gallons. The vessel was then filled with water, and the same quantity again drawn off, and so on for four draughts, when there were only 16 gallons of pure wine left. Required the number of gallons at each draught. *Result*, 27.

(**29.**) A grocer sold 80 lbs. of tea, and 100 lbs. of coffee, for $65. He sold 60 lbs. more of coffee, for $20, than he did tea, for $10. Required the price of each.

*Prices*, $0.50 per lb. for tea; $0.25 per lb. for coffee.

(**30.**) There are two square buildings, that are paved with stones, a foot square each. The side of one building exceeds that of the other by 12 feet, and both their pavements taken together contain 2120 stones. What are the lengths of them separately?

*Ans.*, 26, and 38 feet.

(**31.**) In a parcel which contains 24 coins of silver and copper, each silver coin is worth as many pence as there are copper coins, and each copper coin is worth as many pence as there are silver coins, and the whole is worth 18 shillings. How many are there of each?

*Ans.*, 6 of one, and 18 of the other.

(**32.**) A and B leave at the same time for a place 300 miles distant. A, by travelling 1 mile an hour faster than B, completes his journey 10 hours before him; required their rates of travel per hour.

*Rates*, A 6 miles, and B 5.

(**33.**) A sets out from a certain place, and goes 1 mile the first day, 3 the second, 5 the third, etc. After he has been gone three days, he is pursued by B, who goes 11 miles the first day, 12 the second, etc. When will B overtake A? *Ans.*, 9 days.

(**34.**) What two numbers are those, whose sum multiplied by the greater is 77; and whose difference, multiplied by the lesser, is equal to 12?

Let $xy =$ the greater, and $x =$ the lesser; then by the problem, $x^2y^2 + xy = 77$, and $x^2y - x^2 = 12$;

$\therefore\ x^2 = \dfrac{77}{y^2+y}$, and $x^2 = \dfrac{12}{y-1}$; $\therefore\ \dfrac{12}{y-1} = \dfrac{77}{y^2+y}$,

and clearing of fractions, $12y^2 + 12y = 77y - 77$;

by transposition and division, $y^2 - \dfrac{65}{12}y = -\dfrac{77}{12}$.

*Ans.*, $x = \pm 4,\ \pm \frac{3}{2}\sqrt{2}$; $y = \frac{7}{4},\ \frac{11}{3}$.

(**35.**) A and B gained \$18 in trade. A's money was in use 12 months, and his principal and gain was \$26. B's money, which was \$30, was in trade 16 months. How much did A put in? *Ans.*, \$20.

(**36.**) A person, after doing $\frac{3}{5}$ths of a piece of work in 30 days, calls in another, and they together finish it in 6 days; in what time could each do it alone? *Ans.*, 50, and $21\frac{3}{7}$ days.

(**37.**) A farmer buys $m$ sheep for $p$ dollars, and sells $n$ of them at a gain of 5 per cent.; how ought he to sell the remainder to clear 10 per cent. on the whole? *Ans.*, $p\dfrac{22m - 21n}{20m(m-n)}$, each.

(**38.**) Find two numbers such that their difference shall be 8, and the difference of their 4th powers 14560. *Numbers*, 11, and 3.

(**39.**) A sets out from C to go to D, at the same time that B sets out from D to go to C; A arrives at D, $a$ hours, and B at C, $b$ hours, after they meet. How long was each performing the journey?

*Ans.*, A in $a^{\frac{1}{2}}(a^{\frac{1}{2}} + b^{\frac{1}{2}})$, and B in $b^{\frac{1}{2}}(b^{\frac{1}{2}} + a^{\frac{1}{2}})$.

**(40.)** A certain capital is out at 4 per cent.; if we multiply the number of dollars in the capital by the number of dollars in the interest for 5 months, we obtain \$117041$\frac{2}{3}$. What is the capital?

*Ans.*, \$2650.

**(41.)** What three numbers are they, which, multiplied by two and two, and each product divided by the third number, give the quotients $a$, $b$, $c$? *Ans.*, $\sqrt{ab}$, $\sqrt{ac}$, $\sqrt{bc}$.

**(42.)** If the square of a certain number be taken from 40, and the square root of this difference be increased by 10, and the sum multiplied by 2, and the product divided by the number itself, the quotient will be 4. Required the number. *Number*, 6.

**(43.)** There is a field in the form of a rectangular parallelogram, whose length exceeds the breadth by 16 yards; and it contains 960 square yards. Required the length and breadth.

*Result*, 40 and 24 yards.

**(44.)** A person being asked his age, answered, if you add the square root of it to half of it, and subtract 12, there will remain nothing. Required his age. *Age*, 16.

**(45.)** To find a number from the cube of which, if 19 be subtracted, and the remainder multiplied by that cube, the product shall be 216.

*Number*, 3, or $-2$.

**(46.)** To find a number, from the double of which if you subtract 12, the square of the remainder, *minus* 1, will be 9 times the number sought. *Number*, 11, or $3\frac{1}{4}$.

**(47.)** It is required to divide 20 into two such parts, that the product of the whole and one of the parts, shall be equal to the square of the other. *Parts*, $10\sqrt{5} - 10$, and $30 - 10\sqrt{5}$.

**(48.)** A railway train, after travelling 2 hours, is detained by an accident 1 hour; it then proceeds for the rest of the distance at $\frac{3}{5}$ths of its former rate, and arrives $7\frac{2}{3}$ hours behind time: had the accident occurred 50 miles further on, the train would have arrived $6\frac{1}{3}$ hours behind time. What was the whole distance passed over by the train?

*Ans.*, 300 miles.

**(49.)** Find 6 terms in A. P., whose sum is 45, and whose 1st term is to the 6th as 7 to 8. *Terms*, 7, $7\frac{1}{5}$, $7\frac{2}{5}$, $7\frac{3}{5}$, $7\frac{4}{5}$, 8.

**(50.)** It is required to find two numbers, such, that if their product be added to their sum it shall make 62 ; and if their sum be taken from the sum of their squares, it shall leave 86. *Numbers*, 8, and 6.

(**51.**) The sum of the first two terms of an A. P. is 18, and the sum of the next three is 12. How many terms must be taken to amount to 28? *Ans.*, 7, or 4.

(**52.**) Find the sum of $\sqrt{\frac{3}{2}} + \sqrt{\frac{2}{3}} + \frac{2}{3}\sqrt{\frac{2}{3}} +$, etc., to $n$ terms and to infinity, *Sum*, $\left(\frac{3^n - 2^n}{3^{n-1}}\right)\sqrt{\frac{3}{2}}$ and $3\sqrt{\frac{3}{2}}$.

(**53.**) There are four towns in the order of the letters, A, B, C, D. The difference between the distances, from A to B, and from B to C, is greater by four miles than the distance from B to D. Also the number of miles between B and D is equal to two-thirds of the number between A and C. And the number between A and B is to the number between C and D as seven times the number between B and C : 26. Required the respective distances.

*Distances*, AB = 42, BC = 6, and CD = 26 miles.

---

(**54.**) Find four numbers in arithmetical progression, whose sum shall be 56, and the sum of their squares 864.

(**55.**) There are three numbers in arithmetical progression, and the square of the first added to the product of the other two is 16 ; the square of the second added to the product of the other two is 14. What are the numbers?

(**56.**) The sum of three numbers in arithmetical progression is 9, and the sum of their cubes is 153. What are the numbers?

*Ans.*, 1, 3, and 5.

(**57.**) The sum of three numbers in arithmetical progression is 15 ; and the sum of the squares of the two extremes is 58. What are the numbers?

(**58.**) There are four numbers in arithmetical progression : the sum of the squares of the first two is 34 ; and the sum of the squares of the last two is 130. What are the numbers? *Ans.*, 3, 5, 7, and 9.

(**59.**) A certain number consists of three digits, which are in arithmetical progression ; and the number divided by the sum of its digits is equal to 26 ; but if 198 be added to it, the digits will be inverted. What is the number?

(**60.**) The sum of the squares of the extremes of four numbers in arithmetical progression is 200, and the sum of the squares of the means is 136. What are the numbers?

*Ans.*, $\pm 14$, $\pm 10$, $\pm 6$, $\pm 2$.

(**61.**) The sum of the squares of the extremes of four numbers in arithmetical progression is 200; and the sum of the squares of the means is 136. What are the numbers?

(**62.**) There are four numbers in arithmetical progression whose continued product is 1680, and common difference is 4. What are the numbers? *Ans.*, $\pm 14$, $\pm 10$, $\pm 6$, $\pm 2$.

(**63.**) There are four numbers in arithmetical progression, whose sum is 28, and their continued product 585? What are the numbers?

(**64.**) There are $n$ arithmetic means between 3 and 17, and the last one is three times as great as the first. What is the number of means? *Ans.*, 6.

(**65.**) In any arithmetic progression of which $a$ is the first term, and $2a$ the common difference; prove that the number of terms which must be taken to make a sum $S$, is $\sqrt{\frac{S}{a}}$; $S$ being so assumed that $\frac{S}{a}$ is any square number, but no other.

---

(**66.**) Find three numbers in geometrical progression, whose sum is 14, and the sum of whose squares is 84. *Numbers*, 2, 4, and 8.

(**67.**) There are three numbers in geometrical progression whose product is 64, and the sum of their cubes is 584. What are the numbers?

(**68.**) There are three numbers in geometrical progression; the sum of the first and last is 52, and the square of the mean is 100. What are the numbers? *Ans.*, 2, 10, and 50.

(**69.**) Of four numbers in geometrical progression, the sum of the first two is 15, and the sum of the last two is 60. What are the numbers? *Ans.*, 5, 10, 20, and 40.

(**70.**) A gentleman divided 210 dollars among three servants, in such a manner, that their portions were in geometrical progression; and the first had 90 dollars more than the last. How much had each?

(**71.**) There are three numbers in geometrical progression, the greatest of which exceeds the least by 15; and the difference of the squares of the greatest and the least, is to the sum of the squares of all the three numbers, as 5 to 7. What are the numbers?

*Ans.*, 5, 10, and 20.

(**72.**) There are four numbers in geometrical progression, the second of which is less than the fourth by 24; and the sum of the extremes is to the sum of the means, as 7 to 3. What are the numbers? *Ans.*, 1, 3, 9, 27.

THE END.

# OLNEY'S HIGHER MATHEMATICS.

There is one feature which characterizes this series, so unique and yet so eminently practical, that we feel desirous of calling special attention to it. It is *the facility with which the books can be used for Classes of all Grades, and in Schools of the widest diversity of purpose.* Each volume in the series is so constructed that it may be used with equal ease by the youngest and least disciplined, and by those who in more mature years enter upon the study with more ample preparation. This will be seen most clearly by a reference to the separate volumes.

***Introduction to Algebra***........................

***Complete School Algebra***........................

***University Algebra***........................

***Test Examples in Algebra***........................

***Elements of Geometry.*** Separate........................

***Elements of Trigonometry.*** Separate........................

***Introduction to Geometry.*** Part I. Separate.....

***Geometry and Trigonometry.*** School Edition....

***Geometry and Trigonometry,*** without Tables of Logarithms. University Edition........................

***Geometry and Trigonometry,*** with Tables. University Edition........................

***Tables of Logarithms.*** Flexible covers........................

***Geometry.*** University Edition. Parts I, II, and III...

***General Geometry and Calculus***........................

***Bellows's Trigonometry***........................

---

There is scarcely a College or Normal School in the United States that is not now using some of Prof. Olney's Mathematical works.

They are original and fresh—attractive to both Teacher and Scholar.

Prof. Olney has a very versatile mind, and has succeeded to a wonderful degree in removing the difficulties in the science of Mathematics, and even making this study attractive to the most ordinary scholar. At the same time his books are thorough and comprehensive.

NEW YORK:

SHELDON & COMPANY,

No. 8 MURRAY STREET.

# OLNEY'S SERIES OF ARITHMETICS.

*A Full Common School Course in Two Books.*

OLNEY'S PRIMARY ARITHMETIC, -

OLNEY'S ELEMENTS OF ARITHMETIC,

**A few of the characteristic features of the Primary Arithmetic are:**

*1. Adaptability* to use in our Primary Schools—furnishing models of exercises on every topic, suited to *class exercises* and to pupils' work in their seats.

*2.* It is based upon a *thorough analysis* of the child-mind and of the elements of the Science of Numbers.

*3. Simplicity* of plan and *naturalness* of treatment.

*4. Recognizes the distinction between learning how* to obtain a result and committing that result to memory.

*5. Is full of practical expedients,* helpful both to teacher and pupil.

*6.* Embodies the spirit of the *Kindergarten methods.*

*7. Is beautifully illustrated* by pictures which are *object lessons,* and not *mere* ornaments.

## The Elements of Arithmetic.

*This is a practical treatise on Arithmetic,* furnishing in one book of 308 pages all the arithmetic compatible with a well-balanced common-school course, or necessary to a good general English education.

*The processes usually styled Mental Arithmetic are here assimilated and made the basis of the more formal and mechanical methods called Written Arithmetic.*

Therefore, by the use of this book, from *one-third to one-half the time usually devoted to Arithmetic in our Intermediate, Grammar, and Common Schools can be saved, and better results secured.*

---

These books will both be found *entirely fresh and original in plan,* and in mechanical execution *ahead of any* offered to the public. No expense has been spared to give to Professor Olney's *Series of Mathematics* a dress worthy of their *original and valuable features.*

A Teacher's

## HAND-BOOK OF ARITHMETICAL EXERCISES,

to accompany the ELEMENTS OF ARITHMETIC, is now ready. This book furnishes an *exhaustless mine* from which the teacher can draw for exercise both mental and written in class-room drill, and for extending the range of topics when this is practicable.

## THE SCIENCE OF ARITHMETIC,

The advanced book of the Series, *is a full and complete course for High Schools,* and on an entirely original plan.

SHELDON & COMPANY,

NEW YORK.

# DR. JOSEPH HAVEN'S

# *VALUABLE TEXT-BOOKS.*

Dr. Haven's text-books are the outgrowth of his long experience as a teacher. Prof. Park, of Andover, says of his MENTAL PHILOSOPHY: "It is distinguished for its clearness of style, perspicuity of method, candor of spirit, accuracy and comprehensiveness of thought."

## MENTAL PHILOSOPHY.

1 vol. 12mo. $2.00.

INCLUDING THE INTELLECT, THE SENSIBILITIES, AND THE WILL.

It is believed this work will be found pre-eminently distinguished for the completeness with which it presents the whole subject.

## MORAL PHILOSOPHY.

INCLUDING THEORETICAL AND PRACTICAL ETHICS.

Royal 12mo, cloth, embossed. $1.75.

## HISTORY OF ANCIENT AND MODERN PHILOSOPHY.

Price $2.00.

Dr. Haven was a very able man and a very clear thinker. He was for many years a professor in Amherst College, and also in Chicago University. He possessed the happy faculty of stating the most abstract truth in an attractive and interesting form. His work on "Intellectual Philosophy" has probably had and is having to-day a larger sale than any similar text-book ever published in this country.

**From GEORGE WOODS, LL. D., President Western University of Pennsylvania.**

GENTLEMEN: Dr. Haven's History of Ancient and Modern Philosophy *supplies a great want.* It gives such information on the subject as many students and men, who have not time fully to examine a complete history, need. The material is selected with good judgment, and the work is written in the author's attractive style. I shall recommend its use in this department of study.

**From HOWARD CROSBY, D. D., LL. D., Chancellor of University of New York.**

Messrs. Sheldon & Co. have just issued a very comprehensive and yet brief survey of the History of Philosophic Thought, prepared by the late Dr. Joseph Haven. It is well fitted for a college text-book.

Its divisions are logical, its sketch of each form of philosophy clear and discriminating, and its style as readable as so condensed a work can be. *I know of no compendium which gives the bird's-eye view of the history of philosophy as thoroughly as this hand-book of Dr. Haven.*

SHELDON & COMPANY,

NEW YORK.

# ENGLISH LITERATURE.

## *SHAW'S NEW HISTORY OF ENGLISH AND AMERICAN LITERATURE.*

404 Pages.

Prepared on the basis of Shaw's "Manual of English Literature," by TRUMAN J. BACKUS, of Vassar College, *in large, clear type*, and especially arranged for teaching this subject in Academies and High Schools, with copious references to "The Choice Specimens of English and American Literature." It contains a map of Britain at the close of the sixth century, showing the distribution of its Celtic and Teutonic population; also diagrams intended to aid the student in remembering important classifications of authors.

## *CHOICE SPECIMENS OF AMERICAN LITERATURE AND LITERARY READER.*

518 Pages.

Selected from the works of American authors throughout the country, and designed as a text-book, as well as Literary Reader in advanced schools. By BENJ. N. MARTIN, D. D., L. H. D.

# DR. FRANCIS WAYLAND'S VALUABLE SERIES.

## *INTELLECTUAL PHILOSOPHY (Elements of).*

426 Pages.

By FRANCIS WAYLAND, late President of Brown University.

This work is a standard text-book in Colleges and High Schools.

## *THE ELEMENTS OF MORAL SCIENCE.*

By FRANCIS WAYLAND, D. D., President of Brown University, and Professor of Moral Philosophy.

Fiftieth Thousand. 12mo, cloth,

*** This work has been highly commended by Reviewers, Teachers, and others, and has been adopted as a class-book in most of the collegiate, theological, and academical institutions of the country.

## *ELEMENTS OF POLITICAL ECONOMY.*

BY FRANCIS WAYLAND, D. D., President of Brown University.

Twenty-sixth Thousand. 12mo, cloth,

*** This important work of Dr. Wayland's is fast taking the place of every other text-book on the subject of *Political Economy* in our colleges and higher schools in all parts of the country.

We publish Abridged Editions of both the *Moral Science* and *Political Economy*, for the use of Schools and Academies.

SHELDON & COMPANY,

NEW YORK.

VALUABLE COLLEGE TEXT-BOOKS.

## KENDRICK'S XENOPHON'S ANABASIS.

**533 Pages.**

Comprising the whole work, with Kiepert's Revised Map of the Route of the *Ten Thousand*, Introduction, full though brief Notes, and complete Vocabulary, by A. C. KENDRICK, D. D., LL. D., Rochester University.

## BULLIONS'S LATIN-ENGLISH AND ENGLISH-LATIN DICTIONARY.

(1,258 pages.) This book has peculiar advantages in the distinctness of the marks of the quantities of Syllables, the Etymology and Composition of Words, Classification of Syllables, Synonyms, and Proper Names, and a judicious Abridgment of Quotations. For cheapness and utility it is unequalled.

## LONG'S ATLAS OF CLASSICAL GEOGRAPHY.

This Atlas, by GEORGE LONG, M. A., late Fellow of Trinity College, Cambridge, contains fifty-two Maps and Plans, finely engraved and neatly colored; with a Sketch of Classical Geography, and a full Index of Places. The maps, showing the ideas which the ancients had of the world at various intervals from Homer to Ptolemy, and the typographical plans of ancient places, battles, marches, will be of interest and advantage; and the Atlas will be of great help to classical students, and in libraries of reference.

## BAIRD'S CLASSICAL MANUAL.

(200 pages.) This is a student's hand-book, presenting, in a concise form, an epitome of Ancient Geography, the Mythology, Antiquities, and Chronology of the Greeks and Romans.

## HOOKER'S NEW PHYSIOLOGY.

**376 Pages.**

Revised, corrected, and put into the most perfect form for text-book use, by J. A. SEWALL, M. D,, of the Illinois State Normal University.

***This New Physiology*** has been ***Newly Electrotyped*** in large-sized type, using the ***black-faced type*** to bring out prominently the leading ideas. It contains a full series of ***Questions*** at the end of the book, and a complete ***Glossary*** and ***Index.***

## HOPKINS'S LECTURES ON MORAL SCIENCE.

Delivered before the Lowell Institute, Boston, by MARK HOPKINS, D. D., President of Williams College.

**Royal 12mo, cloth.**

# TEXT-BOOKS ON GOVERNMENT.

## *ALDEN'S CITIZEN'S MANUAL.*

135 Pages.

*A Text-Book on Civil Government, in connection with American Institutions.*

By JOSEPH ALDEN, D. D., LL. D., President of the State Normal School, Albany, N. Y.

This book was prepared for the purpose of presenting the subjects of which it treats in a manner adapted to their study in Common Schools. It has been extensively adopted, and is widely used, with most gratifying results. It is introductory to this author's larger book.

## *THE SCIENCE OF GOVERNMENT,*

**In connection with American Institutions. 295 pages.**

By Dr. ALDEN. Intended as a text-book on the Constitution of the United States for High Schools and Colleges. This book contains in a compact form the facts and principles which every American citizen ought to know. It may be made the basis of a brief or of an extended course of Instruction, as circumstances may require.

# SPELLERS.

## PATTERSON'S COMMON SCHOOL SPELLER.

**160 Pages.**

By CALVIN PATTERSON, Principal Grammar School No. 13, Brooklyn, N. Y.

This book is divided into seven parts, and thoroughly graded.

## PATTERSON'S SPELLER AND ANALYZER.

**176 Pages.**

Designed for the use of higher classes in schools and academies.

This Speller contains a carefully selected list of over 6,000 words, which embrace all such as a graduate of an advanced class should know how to spell. Words seldom if ever used have been carefully excluded. The book teaches as much of the derivation and formation of words as can be learned in the time allotted to Spelling.

## PATTERSON'S BLANK EXERCISE BOOK.

For Written Spelling. Small size. Bound in stiff paper covers.

**40 Pages.**

## PATTERSON'S BLANK EXERCISE BOOK.

For Written Spelling. Large size. Bound in board covers.

**72 Pages.**

Each of these Exercise Books is *ruled, numbered,* and *otherwise arranged* to *correspond with the Spellers. Each book contains directions by which written exercises in Spelling may be reduced to a system.*

There is also an *Appendix*, for *Corrected Words*, which is in a *convenient form* for *reviews.*

*By the use of these Blank Exercise Books a class of four hundred may, in thirty minutes, spell fifty words each, making a total of 20,000 words, and carefully criticise and correct the lesson; each student thereby receiving the benefit of spelling the entire lesson and correcting mistakes.*

www.ingramcontent.com/pod-product-compliance
Lightning Source LLC
LaVergne TN
LVHW021427110826
845150LV00007B/2123

* 9 7 8 1 4 2 5 5 0 7 4 2 8 *